BioMethods
Vol. 2

F. Sierra

A Laboratory Guide to *In Vitro* Transcription

1990
Birkhäuser Verlag
Basel · Boston · Berlin

Author's address:
Dr. Felipe Sierra
Centre de Recherche Nestlé
Nestec S.A.
Vers-chez-les-Blanc
CH–1000 Lausanne 26 / Switzerland

Deutsche Bibliothek Calaloging-in-Publication Data

Sierra, Felipe:
A laboratory guide to in vitro transcription / F. Sierra. – Basel;
Boston ; Berlin : Birkhäuser, 1990
(Biomethods ; Vol. 2)
ISBN 3-7643-2357-4
NE: GT

Library of Congress Cataloging-in-Publication Data

Sierra, F. (Felipe), 1953–
A laboratory guide to in vitro transcription / F. Sierra.
p. cm. – – (BioMethods ; vol. 2)
Includes bibliographical references.
ISBN 0-8176-2357-4 (U.S.)
1. Genetic transcription – – Laboratory manuals. I. Title.
II. Series.
QH450.2.S52 1990
574.87' 3223'078 – – dc20

P. O. Box 133
CH–4010 Basel / Switzerland
Printed in Germany on acid-free paper
ISBN 3-7643-2357-4
ISBN 0-8176-2357-4

Contents

We often discover what will do,
by finding out what will not do;
and probably he who never made a
mistake
never made a discovery.

Samuel Smiles

Acknowledgements

I am grateful to the many colleagues and friends without whose help and encouragement this book would not have been written. I am specially indebted to Professor Ueli Schibler (University of Geneva), in whose laboratory the techniques described were developed. His exceptional scientific and human qualities allowed for innumerable helpful discussions. I would also like to thank Drs. Yves Guigoz (Nestlé Research Center, Nestec, Ltd.), A. Le Cam (CNRS-INSERM, Montpellier), U. Schibler (University of Geneva) and Mrs. R. Walter (ISREC, Epalinges) for helpful discussions and for reading earlier versions of the manuscript. Ms. C. Chaperon, Mr. S. Di Grandi, and Ms. C. Ruffieux helped in checking at the bench that the protocols work, and Mr. M. Juillerat helped with the photography. Finally, I would like to acknowledge the secretarial assistance of Mrs. B. Gehring and the general support and encouragement from people at the Nestlé Research Center.

F. Sierra

I Introduction

A Introductory Remarks

Recent advances in molecular biology techniques have greatly widened our understanding of biological phenomena at the molecular level. Researchers have focused their attention on an array of different aspects of developmental, temporal and structural effects of gene expression. Meanwhile, the control of this genetic activity at the level of transcription remains a fundamental underlying process whose workings are just beginning to be unravelled.

The study of DNA-protein interactions relevant to the problem of eukaryotic gene regulation has recently been made possible through the development of such *in vitro* techniques as *in vitro* footprinting (Galas & Schmitz, 1978), *in vivo* footprinting (Saluz & Jost, 1990), gel retardation (Garner & Revzin, 1981; Fried & Crothers, 1981) and *in vitro* transcription (Manley et al., 1980; Heintz & Roeder, 1982; Dignam et al., 1983; Parker & Topol, 1984a; Gorski et al., 1986; Shapiro et al., 1988). All of these methods depend on the availability of cellular or nuclear extracts that contain, in a soluble form, the factors of interest. One method, developed several years ago (Manley et al., 1980) allows the preparation of such extracts from cells maintained in tissue culture. While a considerable amount of interesting information has been obtained by using this technique, its main drawback has been the difficulty in adapting it to solid tissues derived from the living animal.

The methods I describe and discuss in detail in this book were specifically developed in order to surmount this problem. Most of these techniques have already been published and are therefore available to the scientific community (Gorski et al., 1986; Lichsteiner et al., 1987). Nevertheless, it is clear that many details of manipulation can not be described in regular scientific papers. It therefore ensues that publication of a detailed, step-by-step procedure should be of interest to a large number of scientists. In the same context, I feel that the inclusion of a trouble-shooting guide and some examples of misguided strategies at the end of the book, will significantly shorten the time required to get the protocols working.

Finally, I sincerely hope that by making these detailed protocols available to the scientific community at large, I will encourage their further application, and foster the development of related techniques.

B Technical Approaches to the Phenomenon of Eukaryotic Transcription

In its simplest form, transcription can be regarded as a chemical reaction in which ribonucleotides are polymerized in a defined order determined by a macromolecular template, DNA. The enzyme responsible for catalyzing this reaction is RNA polymerase. Several RNA polymerases have been isolated from mammalian and other eukaryotic organisms (reviewed by Sentenac, 1985; Lewis & Burgess, 1981). Early studies, based on the α-amanitin sensitivity of the various semi-purified enzymes, have already shown that messenger RNA is produced *in vivo* by only one of these isozymes, called RNA polymerase II (pol II). Further studies, making use of defined templates as well as highly purified pol II, clearly demonstrated that pol II, a multi-subunit complex itself, was capable of carrying out the process of transcription on its own, without the requirement of ancillary factors. However, the purified enzyme did not have by itself the capacity to recognize bona fide promoters; instead, these preparations were capable of initiating template-dependent transcription from promoterless regions of DNA (Weil et al., 1979). This observation prompted two interlocking conclusions to be drawn. The first is that the capability to direct pol II to the transcription initiation site is most likely a property of the promoters, rather than of the polymerase. The second conclusion is that, in order for this information to be used by pol II, other factors must necessarily interact with the promoter, the polymerase, or both.

RNA polymerases are fairly conserved throughout evolution (Paule, 1981; Nonet et al., 1987; Saltzman & Weinmann, 1989), and the fact that the same molecule, pol II, carries out mRNA transcription in all the different cells of an organism further strengthens the notion that regulation of gene expression is achieved via specific elements in the promoters, which interact with a cellular machinery that is unique for each cell and for each stage of development.

The best studied examples of regulation of gene activity are derived from bacterial systems, where some of the most important genetic switches have been analyzed by a powerful combination of genetics and biochemistry (Ptashne, 1986). Unfortunately, rapid advancement in our understanding of the analogous phenomena in higher eukaryotes has been partially hampered by the inherent complexity found in such organisms. Nevertheless, many of the basic ideas derived from the bacterial and phage research efforts can now be tested in higher eukaryotic systems.

These studies have been made possible through the advent of powerful techniques, such as genomic cloning (Lawn et al., 1978; Maniatis et al., 1978), *in vitro* mutagenesis (Messing, 1983; Shortle & Botstein, 1983), etc. Probably the single most significant technological breakthrough worth mentioning in this regard is the development of methods for introducing cloned, manipulated genes back into living cells by the method of transfection (Graham & Van der Eb, 1973; Wigler et al., 1979), and later, into the whole organism, resulting in the production of transgenic animals (Hogan et al., 1986). The development of these techniques has allowed molecular biologists to make considerable progress in defining and understanding the basic elements that constitute an eukaryotic transcription unit and play a role in transcriptional control. Among these are enhancers (Serfling et al., 1985), silencers (Borelli et al., 1984; Goodbourn et al., 1986), and the proximal elements that form basic promoters (Maniatis et al., 1987). These DNA motifs (the so-called *cis* elements) have been used in a variety of combinations in order to assess, through transfection and transgenic animal studies, the relative importance of the molecular mechanisms by which they function *in vivo*. While these experiments have clearly shown that *cis* elements work via interaction with nuclear factors (the *trans* elements), the study of these *trans*-acting factors by transfection into living cells or living organisms has been extremely difficult. In fact, transcriptional regulation is mainly the result of interactions between *cis*- and *trans*-acting elements. While we can easily modify the *cis* elements *in vitro*, we can not modify at will the *trans* elements present in the recipient cells or organisms, except for the few instances (such as in yeast and in a handful of higher eukaryotic systems) in which a genetic approach is feasible. In this sense, most of what we can expect to do is to correlate the ability of particular promoter constructions to function in different cells, or to respond to different stimuli. That is precisely the type of experiment that has been performed during the past decade or so, and has given a fairly detailed knowledge of the *cis* elements (Dynan, 1989). In a few cases, carefully designed transfection experiments have allowed a glimpse into the properties of some of the *trans*-acting elements involved. For example, titrations *in vivo* have shown that at least some transcription factors are in limiting amounts in the cell (Schöller & Gruss, 1984), and *in vitro* reconstructed enhancer elements have shown that these are composed of additive, interchangeable modules, each module interacting with a different *trans*-acting factor (Serfling et al., 1985; Dynan, 1989).

Several experimental approaches have been developed in the last few years that allow researchers to investigate the properties and characteristics of the *trans* elements involved in transcription, as well as other processes involving protein-DNA or protein-RNA interactions. As it pertains to the process of transcription, preparation of cellular extracts containing the *trans*-acting factors in a soluble form is a prerequisite for further biochemical characterization. With the advent of several techniques that allow the preparation of such soluble extracts (Manley et al., 1980; Heintz & Roeder, 1982; Dignam et al., 1983; Parker & Topol, 1984a; Gorski et al., 1986; Shapiro et al., 1988), specific protein-DNA interactions became amenable to study, and a series of techniques ensued. The most widely used of these are *in vitro* (Galas & Schmitz, 1978) and *in vivo* footprints (Saluz & Jost, 1990), gel retardation (Garner & Rezvin, 1981; Fried & Crothers, 1981) and filter binding (Fuller & Kornberg, 1983; Fuller et al., 1984). These techniques have allowed considerable insight into the properties of nuclear factors involved in both basal and regulated transcription from a large number of promoters (rev. by Dynan & Tjian, 1985; McKnight & Tjian, 1986; Maniatis et al., 1987; Blau, 1988; Jones et al., 1988; Johnson & McKnight, 1989; Mitchell & Tjian, 1989; Parker, 1989). Sequence-specific DNA-binding activities have in many cases been correlated with a loss of a given function when the corresponding DNA sequence was deleted or mutated, and the manipulated promoter element was then re-introduced into appropriate living cells or animals (Heard et al., 1987). The identification of such potentially interesting protein-DNA interactions has prompted the isolation and cloning of several *trans*-acting factors. This can be achieved by direct screening of expression libraries with the oligonucleotides that interact with the factor in question (Singh et al., 1988; Staudt et al., 1988), or by isolating enough of the natural factor so as to allow micro-sequencing and/or production of antibodies (Young & Davis, 1983; Wiederrecht et al., 1988).

The ultimate proof of function for these purified and cloned transcription factors will necessarily require their re-introduction into appropriate cells or animals, again by transfection and oocyte manipulation. In parallel, studies concerning their structure, function and mode of action can be performed by transcribing the genes they regulate *in vitro*. This assay also provides a rapid and direct means to positively identify potential transcription factors, and the cDNA clones encoding such factors. Even prior to the purification of the proteins, *in vitro* transcription allows direct comparison between the transcriptional activity of a given promoter, and the presence or absence of specific DNA binding proteins in a nuclear extract. The challenge for the future will be to determine the mechanisms by which these protein-DNA interactions can regulate transcriptional activity, and to understand how the whole network of regulatory signals is organized within any given cell. Just as a note of caution, it should be emphasized that many aspects of gene regulation can not at present be analyzed in minute detail. As an example, very little is currently known about the structure of chromatin in general, or about how this structure might affect gene expression. Even theoretically simpler phenomena, such as distance-independent enhancer (or silencer) activity, thought to be largely dependent on protein-DNA interactions, cannot as yet be fully reproduced *in vitro*. It is possible that such a phenomenon might require scaffold attachment sites (Gasser & Laemmli, 1986; Gasser & Laemmli, 1987), and thus, a semi-soluble nuclear extract might be required for properly reproducing it *in vitro*. Further technical developments will thus undoubtedly be forthcoming.

II The Biology of Transcription

A Transcription by RNA Polymerase II

In bacterial systems, the process of transcription is carried out by a single multi-subunit complex, called simply RNA polymerase. Specificity can be achieved by the differential utilization of several sigma factors (Losick & Pero, 1981), as well as activators and repressors. In eukaryotes, however, there are three different RNA polymerases, each one being responsible for transcribing a different type of gene (rev. by Sentenac, 1985). These enzymes can be distinguished from each other by virtue of their sensitivity to inhibition by α-amanitin. Thus, it has been determined that pol I transcribes primarily ribosomal genes, pol II transcribes protein-coding genes, and pol III is responsible for the transcription of tRNA, 5S RNA, U RNA and some other small mRNAs, such as the Adenovirus VA gene (Lewis & Burgess, 1982). The three enzymes do not only share a similar catalytic activity, but also have a substantial structural similarity (Paule, 1981; Saltzman & Weinmann, 1989). Of these three enzymes, pol II gives rise to the RNA class with the highest complexity. This probably requires that the enzyme have a much higher versatility in terms of signals that trigger its activity. These signals are the result of a variety of protein-DNA interactions which occur at the promoter regions of genes transcribed by pol II. Physicochemically, these interactions are however not different from those occurring in pol I and pol III promoters. Nuclear extracts prepared as described in this book contain both pol II and pol III, and probably pol I activities. In this work, however, only mRNA biogenesis will be discussed, which is the result of pol II activity.

B Chromatin versus Promoter Elements

The enormous amount of DNA that each eukaryotic cell contains within its nucleus makes it mandatory for the cell to develop a mechanism for tightly packaging it. To this end, the DNA is associated with histones to form nucleosomes. These structures are then further tightened by the formation of the so-called "higher order structure" of chromatin (Gasser & Laemmli, 1987). Tight packaging in turn has an inhibitory effect on gene expression (Weintraub & Groudine, 1976). This is most likely the result of steric hindrance, which prevents access of the transcriptional machinery to tightly packed regions of DNA. Consistent with this hypothesis, it has been observed that histones, when allowed to form nucleosomes on naked DNA, inhibit initiation and probably elongation of transcription (Workman & Roeder, 1987). However, this does not seem to be always the case, and the subject is still hotly debated. Nevertheless, it has been well documented that genes in an active (or potentially active) configuration are far more sensitive to DNase I digestion than their inactive counterparts (Weintraub & Groudine, 1976). These observations suggest that one early step in gene activation concerns the opening of specific chromatin domains, which then become accessible not only to DNase, but, more importantly, to the transcription machinery. For many genes, chromatin opening appears to be a necessary, but not sufficient, step in gene activation. On the other hand, this step seems to be sufficient for the activation of other genes (Charnay et al., 1984; Sierra et al., 1986; Schibler & Sierra, 1987). It is thus inferred that the latter type of gene only requires an open chromatin structure and ubiquitous transcription factors already present in the cell, while the former genes might require, in addition, the presence of specific transcription factors (Figure 1). Chromatin opening is probably the least understood step in the gene regulation cascade.

For most inducible, as well as tissue-specific or developmentally regulated genes, chromatin opening does not appear to be sufficient for gene activation (rev. by Maniatis et al., 1987). In these cases, a further degree of control is exerted at the level of the promoter. Operationally, we can distinguish two types of *cis* elements which have been identified and shown to control mRNA expression: promoters and enhancers/silencers. Promoters are required for accurate and efficient initiation of transcription. Enhancers and silencers increase or decrease, respectively, the rate of transcription from promoters (Serfling et al., 1985; Dynan, 1989). These

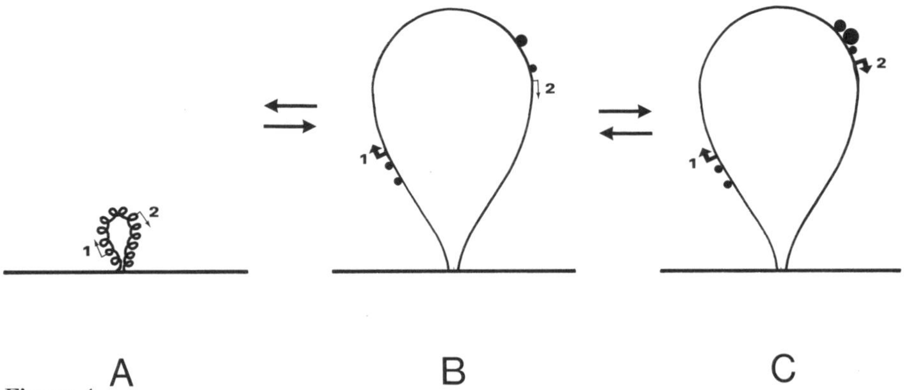

Figure 1.
Chromatin versus Promoter Elements. A hypothetical model for the role of chromatin opening on gene activation. In cells that never express either gene 1 or 2, the corresponding DNA region may be wrapped into non-accessible chromatin (A). Upon opening of the relevant chromatin domain, gene 1, whose promoter only requires ubiquitous transcription factors, becomes active. Binding of ubiquitous transcription factors to the promoter of gene 2 is not sufficient for turning on this promoter (B). Upon appearance within the expressing nuclei of specific transcription factors, gene 2 becomes activated. Since it lacks appropriate binding sites, gene 1 is not affected by the presence or absence of these specific transcription factors (C).

different elements share many properties, but it can be generalized that promoters are located immediately upstream of the start site of transcription, while enhancers and silencers can be located at great distances, upstream or downstream of the promoter, and can generally work independently of orientation (Dynan, 1989).

Apart from the canonical TATA box, promoters usually contain other upstream elements which are important for regulated transcription. Deletion of such elements generally result in a modified level of basal expression, or in a modified responsiveness to extracellular stimuli (Dynan & Tjian, 1985; Yamamoto, 1985; Hamer, 1986). These elements, like enhancers, appear to be composed of different modules, and in some cases modules can be interchanged between promoter and enhancer elements (Bienz & Pelham, 1986; Bohman et al., 1987; Parslow et al., 1987). It should be mentioned that, while most of the promoters studied in detail to date contain a TATA box, this is not the case for house-keeping genes, and several examples of regulated promoters lacking a TATA box also exist (Ayer & Dynan, 1988; Smale & Baltimore, 1989; Kageyama et al., 1989).

While apparently both enhancers (repressors) and promoter elements seem to act at least partially via interaction with soluble nuclear factors, chromatin opening might require the participation of solid phase elements, such as the nuclear matrix or nuclear scaffold. For this reason, nuclear extracts prepared as described in the following protocols are not well suited for studies on the mechanisms of chromatin opening. On the other hand, these extracts are useful for studying protein-DNA interactions relevant to both enhancer and promoter functions. However, functional assays have only been successful in the analysis of promoter elements. The distance and orientation independence of enhancers have not been fully reproduced in *in vitro* transcription experiments using these extracts. It is not yet clear whether this is due to a loss (or inactivation) of crucial factors during the preparation of extracts, or if it is the reflection of as yet poorly understood biological phenomena, such as matrix attachment or DNA conformation. Using a different method for preparing transcriptionally competent extracts, other investigators have reported earlier on the reproduction of enhancer activity *in vitro* (Sassone-Corsi et al., 1984; Sergeant et al., 1984; Schöler & Gruss, 1985).

C *Cis* versus *Trans* Elements

As stated in previous paragraphs, most of the important aspects of gene regulation are the result of specific protein-DNA interactions. Within these interactions, we can distinguish two elements: the DNA sequences in the promoter and enhancer regions are in a linear relationship with the coding region of the gene (even though they are sometimes separated by several Kb). Thus, the DNA portion of these interactions is commonly referred to as the *cis* element. The proteins involved are not necessarily attached to the gene in question, but rather, they can exist in equilibrium with a soluble form. Consequently, the protein portion of the interaction is commonly referred to as the *trans* element (Figure 2).

Both the *cis*- and the *trans*-acting elements can play a role in gene regulation. In the most obvious case, transcription of a given gene in a certain cell type will depend on the availability, within that cell type, of active *trans*-acting elements capable of interacting with a given *cis* element present in its promoter/ enhancer. Conversely, the presence of a

Figure 2.
Cis versus *Trans* Elements. The formation of an active pre-initiation complex (right) depends on the interactions between *cis* elements (DNA sequences directly associated with the relevant promoter) and *trans* elements (proteins that recognize and bind to these sequences). The *cis* element is specific for each promoter, although similar sequences might be shared by a family of promoters controlled coordinately by binding to the same *trans* element. The *trans* elements, on the other hand, are not directly associated with any given promoter, but rather, they exist in equilibrium between the bound and unbound forms. The same *trans* element can bind to several different promoters, and the probability of its binding to any given sequence depends on the relative affinities and the concentration of the *trans* element. Some *trans*-acting elements do not directly bind to the DNA, but instead, they work via protein-protein interactions.

given *trans*-acting factor will only have an effect on the expression of genes which contain a specific DNA sequence in its promoter/enhancer capable of interacting with the protein factor. There is also a growing body of data suggesting that, in many cases, protein-protein interactions play a significant role in gene regulation (Gester & Roeder, 1988).

The degree of interaction can be modulated by modifications in either one of the two partners involved. Thus, for example, the glucocorticoid receptor will not interact with its recognition sequence within the promoter of hormone-regulated genes unless it is allosterically modified by binding to the hormone (Yamamoto, 1985). Other protein factors, such as NF-kB, which binds, among others, to the immunoglobulin promoter region, or HSTF, which activates the heat shock promoter, also exist in the cell in a modified form, that prevents their binding to either DNA or other transcription factors (Sorger & Pelham, 1988; Baeuerle & Baltimore, 1988; Johnson & McKnight, 1989). The DNA sequence can also play a role in gene regulation, since it is also subject to potential modifications. The best studied of these involve methylation of cytosine residues (Bird, 1986; Cedar, 1988). It is known that, in general, high levels of DNA methylation are associated with low transcriptional activity of the promoters affected (Razin & Riggs, 1980). *In vitro*, the interference of methyl groups with protein binding is indeed a very useful technique. In some cases, such as for certain restriction endonucleases, protein-DNA interactions are actually facilitated by the presence of methyl groups in defined positions along the DNA (Nelson & Mc Clelland, 1987). It is conceivable, therefore, that in some cases, methylation of promoter/enhancer sequences might actually induce protein-DNA interactions. In both cases, the effect of methylation on transcription could, at least theoretically, be either positive or negative, depending on the nature (activator or repressor) of the protein binding to the methylation-sensitive site.

D A Case for *In Vitro* Transcription

Nuclear extracts are mixtures of *trans*-acting elements, obtained in a soluble form from the nuclei of the cells of interest. *In vitro*, these *trans* elements are combined with *cis* elements (DNA), and their interactions are measured by means of a variety of techniques. The system can be used to assess the type and concentration of different transcription factors in various extracts; alternatively, the effect of modifications at the DNA level can be measured by using nuclear extracts from only one source. This is exemplified by measurements of *in vitro* transcription rates from a series of deletion or site-specific mutants. In some instances, the effect of methylation at particular sites within the DNA can also be estimated.

It should be emphasized that different methods exist which allow inspection of the various aspects of gene control. Studies of chromatin accessibility are best done *in vivo*, and so, if one wishes to manipulate the elements involved in this process, the use of transgenic animals becomes imperative until new analytical methods are available. Similarly, transfection techniques are particularly well suited for studies concerning the *cis*-acting elements of transcription, but can only provide marginal information about the *trans*-acting elements involved. Soluble nuclear extracts, on the other hand, provide a powerful tool for the study of both *cis*- and *trans*-acting elements. At the same time this system allows the purification and physical characterization of these factors, a prerequisite for understanding the way in which they control gene expression.

Over the years, several methods for preparing cellular extracts have been developed. As early as 10 years ago, Roeder's lab published a method for the preparation of a cytoplasmic extract capable of accurately initiating transcription from viral promoters when supplemented with purified RNA polymerase II (Weil et al., 1979). Since this extract was a cytoplasmic fraction obtained under low ionic strength conditions, it has been widely assumed that the transcription factors contained in it are the result of leaching from the nucleus. Not all macromolecules could efficiently leave the nucleus under these conditions, as evidenced by the dependence of these extracts on the addition of exogenous RNA polymerase II. On the other hand, active RNA polymerase III is found at significantly high levels in these extracts. In 1980, Manley et al. published a protocol, adapted from that of Sudgen & Keller (Sudgen & Keller, 1973), for the preparation of transcriptionally active extracts consisting of a high salt lysate from whole cells (Manley et al., 1980). These extracts contain RNA polymerase II as well as other transcription factors. Whole cell transcription extracts prepared from mammalian (mainly HeLa cells) or *Drosophila* (Parker & Topol, 1984a) tissue culture cells have been widely utilized to study transcription initiation from a variety of cellular and viral promoters. These studies have yielded valuable information regarding general transcription initiation factors, such as the TATA binding factor (Wu, 1985) and other proteins involved in the formation of the pre-initiation complexes (Fire et al., 1984). Several reports have dealt with the presence, within these extracts, of some relatively promoter-specific transcription factors, such as Sp1 (Dynan & Tjian, 1983), HSTF (Parker & Topol, 1984b), and several others (for a review, see Dynan & Tjian, 1985).

The first report on the preparation of transcriptionally active extracts that took advantage of the presumably nuclear localization of the transcriptional machinery also came from Roeder's laboratory (Dignam et al., 1983). In this protocol, nuclei are isolated under low ionic strength conditions by two rounds of low speed centrifugation. The crude nuclei are then lysed by increasing the ionic strength, and extracts are prepared by extensive dialysis and a high speed centrifugation step. This method is now widely used, but, except for a few instances, its main application is restrained to cells maintained in tissue culture. When such cells are available, extracts prepared in this manner are a useful tool for studying both basal and inducible levels of transcription from a variety of cellular as well as viral promoters.

The cell types generally used as a source for extracts are poorly differentiated, and only a few reports have suggested that tissue specificity might be reproduced *in vitro* (Tsuda & Suzuki, 1981; Bazett-Jones et al., 1985; Schöler & Gruss, 1985). For this reason, it was obviously desirable to devise a method for the preparation of nuclear extracts derived from solid tissues obtained directly from the animal. Such a method has been recently developed in the laboratory of Ueli Schibler at the University of Geneva, and the technical details of its preparation are described in the present volume.

III Experimental

A General Considerations

1 Considerations in Developing a Method

The successful preparation of nuclear extracts capable of performing *in vitro* transcription depends on careful manipulation at each step of the protocol. The development of such a method, on the other hand, depends mainly on the careful assessment of theoretical as well as practical steps likely to influence the quality of the end product.
In developing this method, special attention was paid to the quality of the nuclei that are the starting material for the preparation of the soluble extracts. It is clear from the available methods that, during the preparation of nuclei under low ionic strength conditions, considerable amounts of nuclear material are leached into the cytoplasm. Indeed, early successful attempts at preparing transcriptionally active extracts used a cytoplasmic supernatant as starting material (Weil et al., 1979). Even more recent protocols (Dignam et al., 1983) leave behind significant amounts of nuclear material, as evidenced by the capability of the corresponding cytoplasmic fractions to sustain RNA polymerase III-dependent transcription.
In an effort to diminish losses due to nuclear leaching, we prepare the nuclei in an extremely viscous solution. This is attained through the addition of a high concentration of both sucrose and glycerol, while the ionic strength is maintained at a low level (15 mM KCl). While the use of such a solution is technically more cumbersome, we suspect (but have not proven) that the very low diffusion rate characteristic of viscous solutions is quite efficient in diminishing the problem of nuclear leaching.
Frequently, tissues freshly isolated from a live animal contain relatively high concentrations of macromolecule-degrading enzymes, such as DNases, RNases and proteases, as compared to tissue culture cells. Dissociation of solid tissues also induces mechanical stress, which in turn is more likely to induce rupture of lysosomes and other organelles. Several precautions have thus been taken to limit the damage produced by these activities. First, several protease inhibitors (PMSF and Trasylol, a trademark for Aprotinin) are included at all steps prior to the final dialysis. In

the case of particularly difficult tissues, such as pancreas and salivary glands, a more complete cocktail of protease inhibitors is used, and low fat milk is added during the homogenization, as a further way of diminishing loss of activity due to proteases (U. Schibler, personal communication). DNA degradation, on the other hand, could lead to several problems: the presence of excess DNA (coming from the extract, due to inability to eliminate small oligonucleotides) inhibits transcription, as well as footprinting and, most probably, all other assays of protein-DNA interactions. Furthermore, it appears that residual DNA in the preparation can lead to the trapping of histones or other basic proteins, which in turn inhibit transcription. Finally, both the presence of DNA and histones will lead to an erroneous assessment of the concentration of "good" proteins in the extracts. Therefore, an attempt has been made to inhibit DNase activity during the preparation of the extracts. For this purpose, divalent cations such as Mg^{+2} and Ca^{+2} (which are required for DNase I activity) have been replaced by spermine and spermidine. EDTA is included to chelate divalent cations coming from the tissue itself.

Many transcription factors appear to be activated or inactivated by post-translational modifications, such as phosphorylation (Sorger & Pelham, 1988; Bagchi, 1989) or O-glycosylation (Jackson & Tjian, 1988; Lichsteiner & Schibler, 1989). While measures designed to protect such modifications are obviously desirable, none have been included in the protocols presented in this volume. It is therefore quite possible that addition of, for example, phosphatase inhibitors such as NaF, might considerably improve the performance of the extracts with some, although not all promoters.

Finally, no special efforts have been made to inhibit RNases, since they were not judged to be detrimental during the preparation of the extracts. Obviously, they can be quite harmful when present in an *in vitro* transcription reaction; consequently, these reactions are always performed in the presence of a high concentration of RNase inhibitors (we use RNasin, from Promega Biotec), and an excess of tRNA is added as soon as the reaction is finished. When added during the transcription reaction, however, tRNA inhibits transcription from the albumin promoter, although not from the Adeno Major Late promoter.

After isolation of clean nuclei, the preparation of soluble extract follows approximately the method described by Parker & Topol (1984a) for the preparation of whole cell extracts.

2 General Manipulations

The steps in which manipulations are critical or difficult will be discussed in detail along with the relevant protocols. Nevertheless, some general

aspects should be kept in mind throughout the experiments, and these will be discussed in this section.
First of all, the variables of time and temperature can not be overemphasized. Keeping the time as short as possible is crucial for obtaining good quality extracts. This means anticipating the steps ahead, and being prepared. For example, the spectrophotometer should be already prewarmed and set at the correct wavelength before you need to measure the concentration of DNA. Similarly, the ultracentrifuge should be precooled before sacrificing the animals. This way you prevent waiting for those steps, while the sample is "self-degrading" in the ice bucket. Of course, the initial steps (when the preparation is dirtiest) are the most critical, and the time elapsing between killing the animal and starting the first ultracentrifugation should be specially monitored. Perhaps insisting on the necessity of keeping the sample well cooled is unnecessary for people accustomed with biochemistry, but nevertheless, this aspect should also not be underestimated. We find an ice-water mixture to be more efficient at keeping a constant temperature than ice alone.
Again, in some cases it might be advisable to take extra precautions against degrading enzymes such as phosphatases. In all cases, however, the sample should be as protected as possible from the action of nucleases and proteases.
Of course, general laboratory practice should be prevalent when using hazardous chemicals and equipment, including radioactivity. Instructions from equipment manufacturers must be followed in all cases, and particular attention should be paid to local and federal safety regulations.

3 Equipment Required

Most, but not all, of the equipment necessary for the preparation of nuclear extracts and their utilization is standard equipment found in molecular biology or biochemistry laboratories.
The major pieces of equipment are listed below.

3.1 Extract preparation:

- Motor-driven Potter-Elvehjem homogenizer (a home-made version is shown in Figure 4).
- Refrigerated ultracentrifuge
- Rotors: SW 27 and Ti 60 (or Ti 50 if samples are small)
- Vacuum pump fitted with a trap
- Spectrophotometer

- Balance
- Refrigerated Eppendorf-type centrifuge
- Liquid Nitrogen storage space (for long-term storage)

3.2 *In vitro* Transcription and Footprints:

- Radioactivity containment facility
- High power electrophoresis power supply and apparatus
- Gel drier
- X-ray developing facility

Also, the following small pieces of equipment are necessary:

- Surgical equipment
- Glass Dounce homogenizer (with A pestle)
- Dialysis bags
- Water baths
- Timer (seconds)
- Pipetman and mechanical aspirator
- Shaker, vortex and stirrer
- Gel plates, spacers and combs
- Eppendorf and miscellaneous tubes
- Ultracentrifuge tubes

4 Choice of Animals and Tissues

Obviously, the choice of animals and tissues from which to make extracts is dictated by the biological system that interests the investigator. Nevertheless, a general outline of the technical difficulties presented by different tissues might help direct the efforts of those who do have the possibility of choosing.

As for the choice of animals, I have worked almost exclusively with the most common laboratory species: mice and rats. In general, rat liver extracts are more reproducible and, as a direct consequence of the larger size of the animal, yields are a little better than analogous extracts from mice. Even in the case of liver, but more clearly when working with smaller organs, the size of the animal can be important. This does not only have an effect in the yield of extract per animal, but also, dissection is easier (and faster) when using larger animals. This results in less contamination from adjacent tissues. In some cases, tissue specificity is partially achieved through negative regulation in non-expressing tissues. Therefore, tissue contamination can greatly affect the experimental results. For preparative purposes, and when the protein factor can be shown to be evolutionarily conserved, the method has been upscaled to be used with slaughterhouse animals (bovine thymus, for example).

In my experience, the easiest tissue to work with is the liver. It is large, soft, and composed mainly (approximately 70%) of a single cell type, hepatocytes. It expresses many tissue-specific and inducible genes, as well as most metabolic enzymes. Transcriptionally active extracts have nevertheless been prepared from a variety of other tissues, such as spleen, brain and pancreas. Different technical problems have arisen from the use of these different sources of material. Some of these problems have been more or less successfully counteracted by introducing appropriate modifications to the basic protocol.

4.1 Pancreas

The pancreas is also a relatively large, soft tissue. However, it differs sharply from the liver in its content of macromolecule-degrading enzymes. In fact, one of the major functions of this tissue is the secretion of digestive enzymes. Therefore special attention should be paid to eliminate, or at least diminish, these activities from the preparations. This has been largely achieved by the inclusion of more (and more varied) protease inhibitors in the homogenization buffer. It is likely, although I have not tried it, that the addition of 1% low fat milk during the homogenization would also improve these extracts, since this has been shown to be effective for several other tissues (U. Schibler, personal communication). Another potential problem arises from the fact that the pancreas is more heterogeneous than the liver at the cellular level. Indeed, the pancreas can be considered to be two or more intermingled, but separate organs: the exocrine pancreas produces digestive enzymes, while the endocrine (islets of Langerhans) produces a series of different hormones. Tissue-specific transcription can thus be easily masked by the presence of negative regulatory factors coming from the non-expressing portion of the tissue, as well as by inactivation of positive factors during preparation of the extracts (by phosphorylation/dephosphorylation, deglycosylation, etc.).

4.2 Salivary Glands

This is an extremely difficult tissue. Just like the pancreas, its main function is the secretion of digestive enzymes, and extreme care must be exerted to limit their damage. In addition, the problems are compounded by the small size of the glands. This fact forces the use of several animals in order to get enough material, and the serial operation of many animals requires a longer processing time at the first step of the protocol. This results in greater problems with macromolecular degradation. In my hands, extracts from parotid glands have been a complete failure, as they are transcriptionally inactive. However, other authors have successfully prepared transcriptionally active extracts from submaxillary glands, simply by adding a more complex cocktail of protease inhibitors. Parotid extracts, however, also retain enough degrading enzymes to make complementation with active liver extracts unsuccessful. Furthermore, they fail to perform even in the simplest assays for protein-DNA interactions: DNases present in the extracts degrade the probes used in gel retardation experiments. This effect can be overcome by the chelation of Mg^{2+} ions through the addition of EDTA. Similarly, DNase footprint assays might be performed with DNase II, which cuts the probe less frequently (making results more difficult to interpret), but does not require Mg^{2+} for its activity. All of these experiments were performed with extracts prepared in the same manner as for liver extracts, and no further attempts have been made to incorporate tricks developed for use with pancreas, spleen or submaxillary gland extracts, such as the addition of extra protease inhibitors and/or low fat milk. Therefore, it is possible, and even likely, that good parotid extracts are feasible, provided a little extra care is taken during their preparation.

4.3 Spleen

Transcriptionally active nuclear extracts from rat spleen are easily prepared, provided a few precautions are taken. Spleen cells are rather small, with very little cytoplasm. Relative to other tissues, they contain less nuclear proteins per ug of DNA. Therefore, extracts should be made twice to three fold as concentrated with respect to the DNA measurement. Also, for reasons that are not understood, spleen extracts are frequently less active in transcription assays than liver extracts. This problem can be fully avoided by inhibiting protease activity through the inclusion of 1% low fat milk in the homogenization buffer.

4.4 Brain

Transcriptionally active nuclear extacts from rat brain are also easily prepared. Inclusion of 1% low fat milk in the homogenization buffer is recommended. Other modifications to the basic protocol have been described (Tamura et al., 1989).

4.5 Age Effects

In my most recent work, I have been interested in the phenomenon of aging. In preparing liver nuclear extracts from aged animals, an unexpected problem arises. Probably due to the high level of polyploidy present in the liver of aged animals, the DNA:protein ratio in isolated nuclei is unusually high. Transcriptionally active extracts have been prepared from this source simply by considering that the nuclear protein to DNA ratio is only 1/4 to 1/3 of that found in younger animals. Also, when working with very young animals (before weaning), it is difficult to prepare transcriptionally active extracts by the standard method. Inclusion of low fat milk in the homogenization buffer completely abolishes the problem (J. Wuarin & U. Schibler, personal communication).

4.6 Tissue Culture

In general, the method described in this volume is not likely to be the best choice for the preparation of transcriptionally active extracts from tissue culture cells, since individual cells do not easily break in 2 M sucrose. I would recommend the use of the method described by Shapiro et al. (1988).

5 Flow Diagram

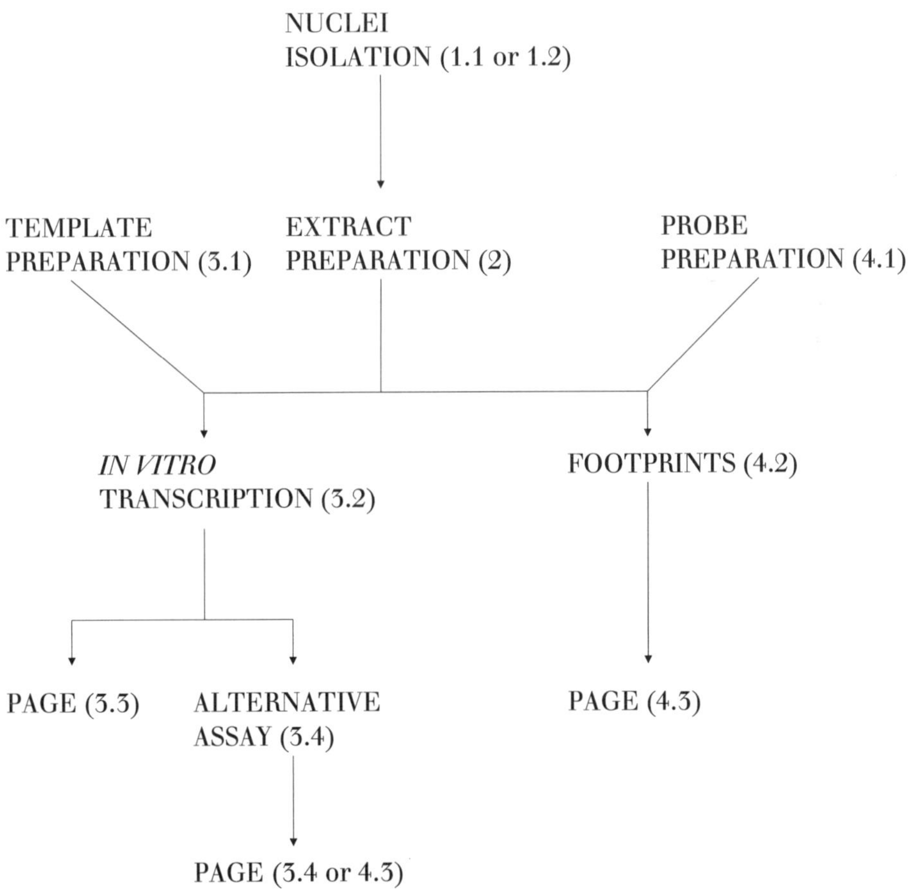

B Methods

1 Nuclei Isolation

This is the most critical step in the preparation of transcriptionally active extracts. The integrity and purity of the nuclei is essential, and the major considerations that must be kept in mind at this step have been discussed in sections III.A.1, and III.A.2. The use of freshly isolated tissue is imperative, since it has been found that extracts prepared from frozen samples are transcriptionally inactive (U. Schibler, personal communication).

Operationally, this protocol can be divided into three steps: dissection, homogenization and centrifugation. However, working fast is crucial at this point, and therefore, these steps are given as a single protocol.

Two alternative protocols (1.1 and 1.2) are given. The first is the standard protocol for preparation of transcriptionally active extracts. At the end of this protocol, I give several possible modifications that have been designed for preparing extracts from different tissues. The second protocol includes a few shortcuts designed for the preparation of large quantities of extract, usually for preparative protein purifications. Extracts prepared by method 1.2 are usually, but not always, active in transcription assays.

The time required to perform this protocol varies greatly, depending on the tissue to be used (hence, the number of animals and the difficulty of dissection) and also on the number of centrifugations required (one or two). In general, it has been found that nuclei are fairly clean already after the first centrifugation, provided that the walls of the tube are thoroughly rinsed. When in doubt, however, it is advisable to go ahead with the second spin, as this is likely to save you time in the long run. For liver extracts, and with two ultracentrifugation spins, it usually takes approximately 3 to 4 hours to isolate the nuclei (a morning), and these can either be frozen or used immediately to prepare extracts (Protocol 2) in the afternoon.

Materials and Buffers

- Chamber for anesthesia
- Ether
- Dissection equipment: scissors and tweezers
- Squirting bottle with ethanol
- Small beaker
- Two ice buckets
- Motor-driven homogenizer (Teflon-tipped Potter-Elvehjem)
- Large plastic beaker with ice-water (500–800 ml)
- Ultracentrifuge, pre-cooled to –2°C
- SW 27 or equivalent rotor (pre-cooled) and appropriate tubes
- Water pump or other suction equipment, with a trap for liquids
- Large spatula
- Mechanical pipettor
- Plastic 10-ml pipets
- 50- or 100-ml syringe, fitted with long, L-shaped needle
- Glycerol, only if two centrifugations are necessary
- Dry ice, only if nuclei are to be frozen

Buffers

Homogenization Buffer

	For 400 ml
10 mM HEPES, pH 7.6	4 ml, 1 M
15 mM KCl	2 ml, 3 M
0.15 mM Spermine	600 µl, 0.1 M
0.5 mM Spermidine	200 µl, 1 M
1 mM EDTA	800 µl, 0.5 M
2 M Sucrose	273.8 g
10% Glycerol	46 ml, 87%

Just before use, add:

0.5 mM DTT (50 µl of 1 M stock for 100 ml)
0.5 mM PMSF (500 µl of 0.1 M stock for 100 ml, in isopropanol)
1% Trasylol (1 ml of 100% for 100 ml)

Notes:

- To prepare homogenization buffer, weigh out the sucrose first, then add just enough water to wet it, add the glycerol and complete almost to volume with water. Stir in warm room (37°C) until dissolved, then add the other components. Store at –20°C at least overnight. This solution should be used very cold (not warmer than –10°C), because the lower temperatures help to increase its viscosity.
- This buffer can also be prepared with 2.2 M sucrose and only 5% glycerol. This way, a slightly larger amount of tissue can be processed (1.5 to 2 g/10 ml).
- PMSF (phenyl-methyl-sulfonyl-fluoride) is a Serine-protease inhibitor. It is soluble in isopropanol at room temperature. For storage, the solution is kept at –20°C, where the PMSF crystallizes. It can be redissolved by a brief heating at 37°C, and kept at room temperature until needed; if kept on ice, it will re-crystallize! It is important to add this product to the aqueous solution just before use, since the half life of PMSF in water is very short (approximately half an hour).
- Trasylol is a trademark (Bayer, AG) for Aprotinin, sold as a solution at 1.4 mg protein/ml.

20% Low Fat Milk

20 g of powdered low fat milk in 100 ml of H_2O. Spin at 8,000 rpm for 20 minutes. Discard the pellet and add 1/20 of this supernatant to the homogenization buffer.

Notes:

- The 20% low fat milk solution should be made freshly each time.
- I do not add low fat milk to the buffer when preparing regular liver extracts. However, its use is recommended for all other tissues, specially spleen, pancreas, salivary glands, etc. It has also recently been found that low fat milk addition is beneficial during the preparation of liver nuclear extacts from very young animals (less

than three weeks of age (J. Wuarin & U. Schibler, personal communication).

– Low fat milk should never be included in the cushions.

Step-by-Step Procedure

1.1 Regular Nuclei Isolation

- > Before starting, make sure that the homogenization buffer is not partially solidified. If that is the case, warm it briefly at 37°C and then return it to –20°C until needed.
- > Turn on the ultracentrifuge, so that the temperature equilibrates at –2°C (this requires the vacuum to be on).
- > Aliquot the amount of homogenization buffer required for the homogenization and first centrifugation step. This depends on the number of SW27 tubes to be used, as well as the volume of tissue to be processed (see below). Each SW27 tube holds approximately 40 ml (6 tubes per rotor= 240 ml).
- > Add the DTT and Trasylol (but not the PMSF) to this buffer. Due to the viscosity of the solution, this requires vigorously shaking it, until no Schlieren lines are obvious.
- > Add 10 ml of this buffer to each SW27 tube, to form the pads. Keep these on ice at all times.
- > Place the Potter-Elvehjem homogenizer in the large plastic beaker with ice-water, and add a few milliliters of homogenization buffer at the bottom of the homogenizer (this is to prevent the tissue from getting stuck at the bottom).
- > At this point, you can start the dissection step, which is done differently according to the tissue. In general, I prefer to anesthetize the animals with ether, but this, of course, depends on personal experience. It is important to clean the skin of the animal with the squirt bottle containing

ethanol, since this will reduce contamination with hairs, which can damage the homogenizer.

> While the animals are being anesthetized, add PMSF to the homogenization buffer and mix well as before. Also, add 50 µl of PMSF to each of the pads in the SW27 tubes.

> Dissect the tissue and keep it in a cold beaker on ice until all animals have been dissected. Mince as finely as possible with scissors.

> Homogenize the tissue by 3 or 4 strokes in the Potter-Elvehjem homogenizer. For this, the

Figure 3.
Homogenization Set-up. The homogenization step requires a motor-driven conical Teflon pestle. The tube containing the sample should be kept on ice-water at all times. Unless otherwise provided by the manufacturer, the ice-water container should be plastic, with a fairly resistant bottom, so that it can withstand some pressure. It is convenient, though not absolutely necessary, to have a mechanical means for up and down motion (either the motor and pestle, or the sample container). Notice that the amount of homogenization buffer in the container should not be such that the pestle will displace the tissue outside of its own working space.

tissue is resuspended in homogenization buffer at about 1 g per 10 ml:

60 ml per rat liver (young)
150 ml per rat liver (old)
6 ml per rat pancreas
10 ml per rat spleen
3 ml per rat parotid
10 ml per rat thymus

The actual volume used depends also on the number of SW27 tubes to be used at the next step. The tissue should never constitute more than 10 to 15% of the total volume, since this would decrease the actual viscosity of the homogenate, potentially resulting in nuclear leaching.

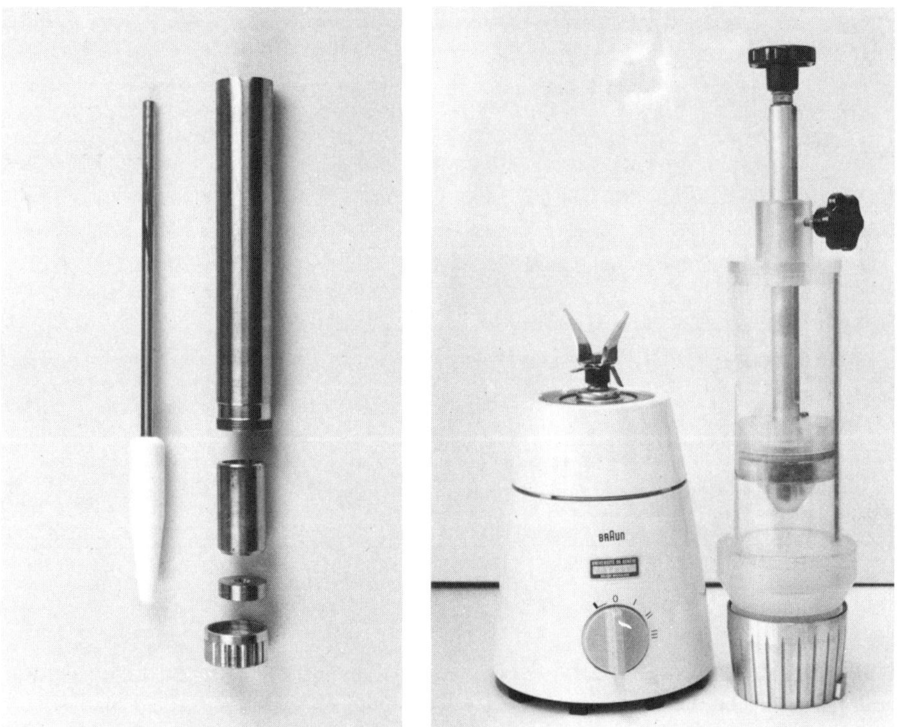

Figure 4.
The "Bazooka" and the Modified Food Processor. Home-made homogenizers used for regular extract preparations ("bazooka", left) or for large-scale preparative assays (modified food processor, right). The blueprints for construction of these are given in Figures 32 and 33, respectively.

The homogenate should be kept on ice-water at all steps, including the homogenization itself. This is a step that requires a certain physical strength, unless mechanical devices are used. Due to the strong shearing forces that develop in the highly viscous sucrose solution, we have occasionally had implosions (but never explosions), resulting in the loss of both the homogenizer and the homogenate. For this reason, it is preferable to use a home-made stainless steel homogenizer with a Teflon pestle (called "bazooka", Figures 4 and 32).

> After most of the tissue has been homogenized, load it on top of the 10-ml pads in the SW27 tubes. No special precaution should be necessary at this step, since the two solutions do not easily mix, due to the dilution of the homogenate by the tissue. Each tube should be loaded to the top (28 ml of extract, 38 ml volume with the pad).

> Repeat the homogenization and loading of the tubes until all animals are processed (or all tubes are filled). Try to do these steps as fast as possible, even if that means being a little sloppy with the actual volumes or with the working space, which can be cleaned after!

> Spin at 24,000 rpm for 60 minutes in a pre-cooled centrifuge (–2°C).

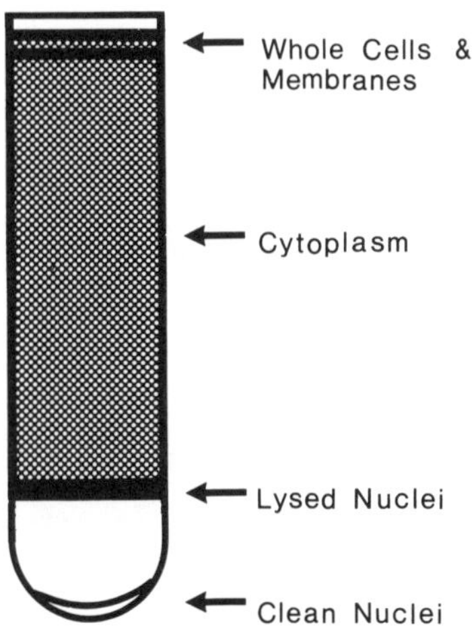

Figure 5.
Appearance of the Ultracentrifuge Tube After Spinning. The diagram shows the approximate appearance of the tubes, as well as the localization of several cellular and subcellular components.

Whole cells and unbroken pieces of tissue float.

Nuclei with cytoplasmic tags and broken nuclei stay at the interface.

Clean nuclei pellet.

> While spinning, check a small aliquot of material under the microscope. Most cells must be broken, but most nuclei should be intact.

The nuclear pellet should be transparent or slightly white, but not reddish, as this indicates contamination with red blood cells or other subcellular debris. Should this be the case, a second centrifugation step is imperative.

> With a spatula, remove the solid disc that floats, then carefully aspirate off the liquid (with 5- or 10-ml pipet, but not with Pasteur pipet), wipe off the sides of the tube with a Kleenex and set the tube on an ice bucket, turned slightly upside down, so that excess buffer runs towards the outside, and not towards the nuclei. Nevertheless, be sure the nuclear pellet is still protected by ice (see Figure 6).

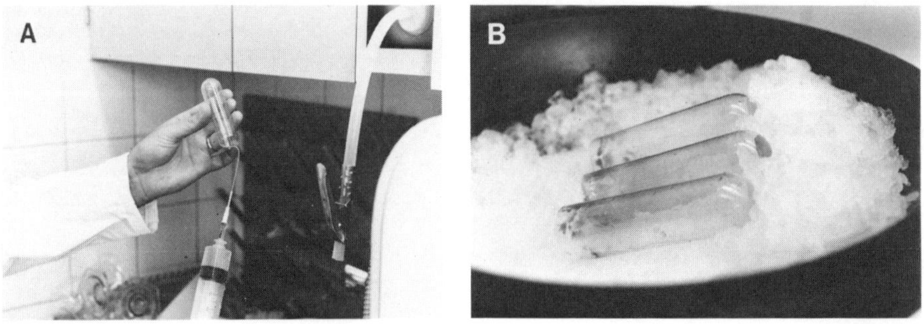

Figure 6.
Washing and Temporary Storage of the Tubes. A. After removal of most of the liquid in the tube by aspiration, the walls are thoroughly washed with distilled water, applied using a large caliber syringe fitted with a long needle that has been bent at a 90° angle at the tip. B. Both before and after the washing just described, the tubes are placed slanted downwards on an ice bed, carefully assuring that ice is protecting the nuclear pellet at all times.

> Fill a large syringe with distilled water and attach to it a long needle with the tip bent about 90°. Holding the tubes upside down over a sink, thoroughly rinse the walls by squirting water with the syringe. Be careful not to wet the nuclear pellet with water, as this would induce lysis. Carefully wipe off the excess water with a Kleenex.

> If the nuclei are clean, they can be frozen by putting the tube directly in dry ice, and kept at –70°C for several weeks, or you can proceed immediately to prepare the extract (Protocol 2).

> If a second centrifugation is deemed necessary, the nuclei should be resuspended in complete

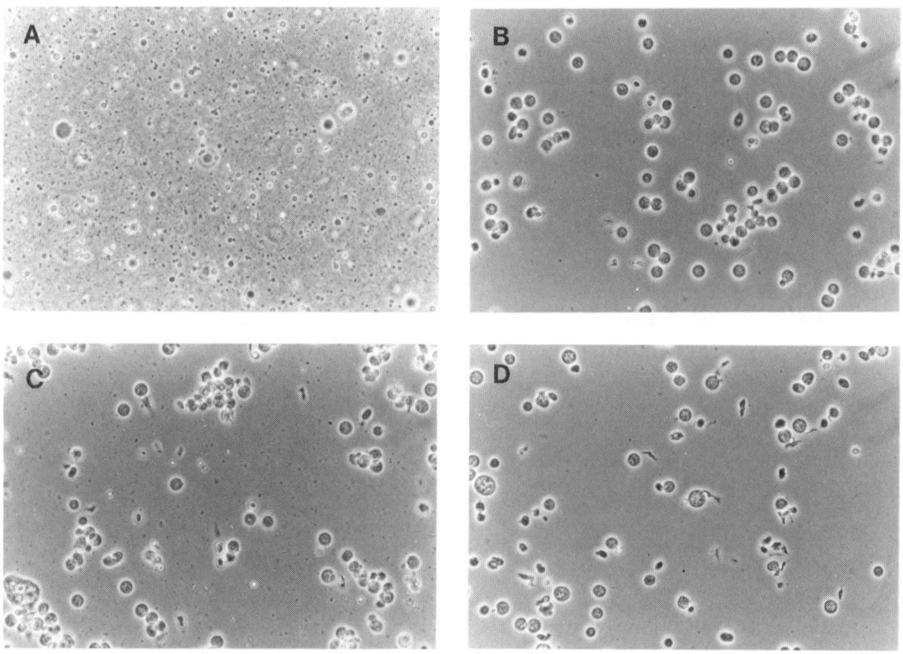

Figure 7.
Nuclei at Different Stages During Preparation. A. Total homogenate, before ultracentrifugation. B. Clean nuclei obtained after a single ultracentrifugation through a sucrose cushion. C. Nuclei are contaminated by red blood cells and other debris after the first centrifugation, due to an excessive amount of tissue being processed at one time. D. Nuclei are contaminated by cytoplasmic tags due to insufficient homogenization.

homogenization buffer containing an additional 10% of glycerol. Usually, this step is carried out in about 50 ml (two SW27 tubes). The nuclei are resuspended by bubbling and scraping the side of the tube with a plastic pipet (avoid glass pipets, as nuclei stick much more to them), then, carefully aspirate up and down several times, making sure not to produce foaming.

- Transfer to a Potter-Elvehjem and homogenize again, only once or twice this time, and with the motor set at a lower speed.
- Layer again onto a 10-ml cushion in SW27 tubes, and centrifuge as before. Clean the tubes as just described, and then freeze in dry ice or continue directly with Protocol 2.
- Check the nuclei again under the microscope.

Modifications Applicable to Specific Tissues

As previously mentioned, in many cases, the addition of extra protease inhibitors and/or low fat milk is desirable.

A more complete protease inhibitor cocktail has been described(Cockell et al., 1989). This cocktail is used in addition to the standard PMSF and Trasylol:

Inhibitor	Stock	
Leupeptin Pepstatin Antipain	0.5 mg/ml each in ethanol	Add 100 μl of each stock for 500 ml of homogenization buffer
TLCK TPCK	10 mM stock in H_2O	

This same cocktail is used during the extract preparation (Protocol 2).

When using low fat milk, it should be used at a concentration of 1% in the homogenization buffer. Do not add the powder directly to the buffer; the preparation of a stock solution is described under Materials and Buffers.

Finally, it is important that low fat milk not be included in the cushions used during centrifugation. In this way, no materials coming from the milk can contaminate the final preparation.

In some cases, when working with small organs, it is better to add less material per volume of homogenization buffer, but to spin only once through a cushion. This way it is possible to diminish losses, and there are usually no problems of reproducibility. It should be mentioned that not all tissues have the same ratio of DNA/nuclear proteins (for example, nuclei from spleen and livers from old animals have less protein) and not all tissues have the same number of nuclei per gram of tissue. Thus, the yield of nuclear proteins from tissues such as kidney or salivary glands is about 1/5 of that from young liver, even though the DNA/protein ratios are similar.

1.2 Large-Scale Nuclei Isolation

In many cases, such as for the preparative isolation of specific transcription factors, the protocol described above for the preparation of nuclei becomes rate-limiting, and a few modifications have been made to speed it up. Some of these modifications can be deleterious for certain specific purposes, so it is up to the investigator to decide which, if any, of these shortcuts are applicable in each case.

The protocol used for large-scale preparations is very similar to the one just described, but with the following modifications:

- The concentration of sucrose is increased to 2.4 M. This allows a larger volume of tissue to be processed per ml of homogenate. Up to 3 g/10 ml of buffer can be homogenized.
- Conversely, no glycerol is used in this buffer, so that the sucrose can go into solution.
- Low fat milk is added at a concentration of 1%. This limits proteolytic damage during the preparation. No milk should be added to the sucrose cushions. See Protocol 1.1 for the preparation of low fat milk stock.
- Homogenization is done in a modified food processor designed in U. Schibler's laboratory (Figures 4 and 33). This allows for faster and easier manipulations at this critical step.
- Unfortunately, it has not been possible to avoid the use of the ultracentrifuge, and the relatively low capacity of the tubes makes this step the most time consuming. Nevertheless, for preparative purposes, spinning just once, and for only 40 minutes seems to be adequate, since further chromatographic purifications will be introduced later. Also, 7.5-ml cushions can be used, so that a little more sample can be loaded per SW27 tube.

In summary, the composition of the homogenization buffer used for this protocol is as follows:

Homogenization Buffer (preparative)

	For 1 liter
10 mM HEPES, pH 7.6	10 ml, 1 M
15 mM KCl	5 ml, 3 M
0.15 mM Spermine	1.5 ml, 0.1 M
0.5 mM Spermidine	500 μl, 1 M
1 mM EDTA	2 ml, 0.5 M
2.4 M Sucrose	821.5 g
1% low fat milk	50 ml, 20 %

Just before use, add 0.5 mM DTT, 0.5 mM PMSF and 1% Trasylol.

Step-by-Step Procedure

1.2 Large-scale Nuclei Isolation

> Complete the homogenization buffer and precool the ultracentrifuge as for protocol 1.1.

> Prepare the pads in the SW27 tubes, but use only 7.5 ml per pad. The pads should be done as before, that is, with regular homogenization buffer containing only 2 M sucrose, 10% glycerol and no low fat milk. Nuclei will not properly pellet through a 2.4 M sucrose pad.

> Proceed with the dissection as before, but do not mince the tissue. Up to 60 g of tissue can be processed per SW27 rotor (in about 200 ml = 0.3 g/ml).

> Put the appropriate amount of buffer in the modified food processor (Figures 4 and 33). Put over this a thin layer of the same buffer, but less cold (4°C). This prevents the tissue from freezing on contact.

> Add the tissue and close the apparatus carefully, so as to avoid any air bubbles. This is important, since air bubbles will induce the formation of foam, which denatures proteins, as well as enlarging the volume of the prep.

> Homogenize in a short burst (8 to 15 s should be enough). As fast as possible, remove the lid and pour the homogenate into a shallow container kept in ice-water (the temperature of the homogenate rises to 8–10°C during this short manipulation).

> Pour the homogenate over the sucrose pads and spin at 24,000 rpm for 40 minutes.

> Remove the supernatant and rinse the walls of the tube with distilled water as in protocol 1.1.

> Freeze in dry ice and store at –70°C or, if possible, under liquid N_2.

> Proceed to step 2 (extract preparation) only after accumulating enough nuclei for the intended purpose.

2 Extract Preparation

This part of the protocol is very similar to the one used by Parker & Topol (1984a). Briefly, it involves resuspending the nuclei and lysing them in a low concentration of $(NH_4)_2SO_4$. After centrifugation to eliminate chromatin, soluble proteins are precipitated by increasing the salt concentration (salting out). Finally, the excess of salt is removed by extensive dialysis. Proteins that precipitate during this step (histones?) are eliminated by a brief centrifugation.
The time required for this protocol is about 7 hours, considering that the last four are spent on dialysis, with only one change of buffer in between. The precipitated proteins can be stored overnight in ice-water before resuspending in dialysis buffer; this cuts the time into two periods of 3 to 4 hours each.

Materials and Buffers

- > Spectrophotometer, pre-set at 260 nm
- > Quartz cuvettes
- > Ultracentrifuge, pre-cooled to –2°C
- > Rotors and tubes (Ti 60 or Ti 50 rotors and tubes)
- > Mechanical pipettor
- > Plastic 10-ml pipets
- > Glass homogenizer fitted with an A pestle
- > Vortex and some small (about 5-ml) tubes
- > 0.5% SDS
- > Shaker (in the cold room)
- > Ice bucket
- > Balance
- > Solid ammonium sulfate
- > Large container for dialysis (if possible, in the cold room)
- > Stirrer and magnet
- > Dialysis tubing and clamps
- > Cold Eppendorf centrifuge
- > Screw cap tubes (1–1.5 ml) for liquid N_2 storage
- > Dewar flask with liquid N_2
- > Long-term liquid N_2 storage space

Buffers

Nuclear Lysis Buffer

	For 100 ml
10 mM HEPES, pH 7.6	1 ml, 1 M
100 mM KCl	3.33 ml, 3 M
0.1 mM EDTA	20 µl, 0.5 M
10% glycerol	11.4 ml, 87%
3 mM $MgCl_2$	300 µl, 1 M

Just before use, add:

1 mM DTT	(100 µl of 1 M stock for 100 ml)
0.1 mM PMSF	(100 µl of 0.1 M stock for 100 ml)
1% Trasylol	(1 ml of 100% for 100 ml)

For tissues other than liver, add the protease inhibitor cocktail described in Protocol 1.1 (see "Modifications applicable to other tissues").

Dialysis Buffer

	For 1 liter
25 mM HEPES, pH 7.6	25 ml, 1 M
0.1 mM EDTA	200 µl, 0.5 M
40 mM KCl	13.3 ml, 3 M
10% glycerol	114 ml, 87%
1 mM DTT	1 ml, 1 M

4 M $(NH_4)_2SO_4$

52.85 g for 100 ml. Adjust to pH 7.9 with NaOH

Notes:

- The same precautions and notes concerning PMSF and Trasylol given in protocol 1.1 are valid for this protocol.
- Nuclear lysis buffer should be stored frozen at –20°C.
- Dialysis buffer is made fresh each time. It should be prepared well enough in advance so as to allow it to cool to 4°C before use.
- Ammonium sulfate is stored either at 4°C or at –20°C. It crystallizes, so it should be stirred for 1 or 2 hours before use (can be warmed to 37°C to re-dissolve it). This solution has occasionally been replaced by 3 M KCl, but the results are less reproducible.

Step-by-Step Procedure

Extract Preparation

> Resuspend the nuclei in completed lysis buffer, again by bubbling and scraping with a plastic 10-ml pipet (do not use glass pipets to handle the nuclei).
> The volume of lysis buffer used depends on the tissue:
>
> 5 ml per rat liver (young)
> 2.5 ml per rat liver (old)
> 1 ml per rat pancreas
> 5 ml per rat spleen
> 100 μl per rat parotid
> 5 ml per rat thymus

> Homogenize the nuclei by several strokes with a hand-held glass homogenizer fitted with an A pestle.

> Measure A_{260} of a 1:50 dilution in 0.5% SDS (blank against a 1:50 dilution of lysis buffer). In order to obtain a homogeneous lysate at this step, the sample is added dropwise into the SDS solution, while vortexing vigorously. Keep vortexing for about a minute after addition of the nuclei.

> From the A_{260}, calculate the concentration of DNA (1 A_{260} = 50 μg/ml).

> Dilute the sample with lysis buffer so as to have a DNA concentration of:
> 0.5 mg/ml for most tissues
> 0.25 mg/ml for spleen and thymus

> Add 1/10 volume of 4 M $(NH_4)_2SO_4$, and mix gently by inversion several times.

In general, it is better to calculate the amount of $(NH_4)_2SO_4$, mix that with the amount of lysis buffer required to dilute the sample (previous step), and then add that mix to the concentrated nuclei. This way there is less possibilities of salting out proteins due to locally high salt concentrations.

Alternatively, this step can be carried out by adding 3 M KCl to a final concentration of 0.55 M (taking into consideration that the lysis buffer already contains 0.1 M KCl). I find this method less reliable, though.

> Leave in ice-water for 30 minutes, with occasional mixing.

> Spin for 60 minutes at 35,000 rpm in the Ti 60 rotor (or at 40,000 rpm in the Ti 50 rotor). If KCl was used as a salt in the previous step, spinning for only 30 minutes might be adequate.

At this step, it is important to remove and process the tubes as soon as the centrifuge stops, since otherwise the chromatin will start re-swelling.

> Transfer the supernatant to a new Ti 60 or Ti 50 tube. This is done by very carefully pipetting the liquid out. Be careful to avoid any contamination with pelleted chromatin, which is transparent and gelatinous. A noticeable disturbance of the chromatin during the pipetting indicates that it is being pulled by the pipet. Stop pipetting immediately and measure the volume of supernatant.

> Add 0.3 g of solid $(NH_4)_2SO_4$ per ml of supernatant.

> After all the salt has dissolved, leave in ice-water, shaking gently, for 20–60 minutes.

> Spin for 20 minutes at 35,000 rpm in the Ti 60 rotor (or 40,000 rpm in the Ti 50 rotor).

> Aspirate off all the supernatant and discard it. Mark on the tube the position of the pellet (it will become very transparent as you start resuspending it).

At this point, the pellet can be stored overnight in ice-water.

> Resuspend in dialysis buffer by gently pipetting

up and down, but being careful to avoid foaming. The volume depends on the DNA measurement done at the beginning of the protocol. The extract should be resuspended at a concentration of approximately 10 mg/ml. Usually, 1 mg of liver nuclear proteins can be extracted per mg of DNA. However, for other tissues, like spleen, thymus or liver from old rats, which contain less soluble nuclear proteins than young liver, the pellets have to be dissolved in two- to four-fold smaller volumes as compared to liver, in order to obtain extracts containing similar concentration of nuclear proteins.

When the dialysis buffer is added, the pellet becomes transparent, giving the false impression of being completely dissolved. Continue resuspending for approximately 30 to 60 minutes, in order to ensure complete recovery of the pelleted proteins. Avoid foaming, as this denatures proteins.

- Dialyze 2x2 hours in the cold against 100 volumes each time of dialysis buffer. A white precipitate should form.
- Remove the dialysate and spin 2 minutes in a cold Eppendorf centrifuge.
- Keep the clear supernatant. Aliquot and freeze immediately in liquid N_2.
- Use a small aliquot of the final extract to determine the protein concentration as follows: measure A_{230} and A_{260} of a 1:50 dilution of extract, made in H_2O, and blanked against a 1:50 dilution of dialysis buffer.

Calculate the protein concentration:

$187 \times A_{230} - 81.7 \times A_{260} = \mu g/ml$
(from Kalb & Barnlohr, 1977).

A good extract should give a concentration of protein between 6 and 10 mg/ml, with a ratio of A_{230}/A_{260} of 4 to 6. This means readings in the range between 0.7 and 1.2 at 230 nm, and corresponding readings approximately 1/5 of those values at 260 nm.

3 *In Vitro* Transcription

The quality of the extract can be measured by a variety of different functional tests, such as *in vitro* transcription, footprinting, gel retardations, etc. Of these, *in vitro* transcription is the most demanding and complex. If the extract is capable of a good level of *in vitro* transcription, then it will certainly perform adequately in the other assays. This reaction requires a complex interaction between the template, several transcription factors, and RNA polymerase.

In principle, there are several criteria that should be tested, such as the concentration of both extract and template that give optimal transcription, inhibition of transcription by different concentrations of α-amanitin, time course of the reaction, temperature, etc. In some cases, it has been found that specific promoters work best under slightly different ionic conditions than others. Since measuring all of these parameters could be rather time-consuming, I routinely test each extract only for the optimal extract concentration. All other parameters are kept constant at the levels found to be optimal when the method was first developed, some of which are described in detail in section III.C.

As a standard template, I use a DNA mixture consisting of a plasmid containing a promoter active in all tissues, such as Adeno Major Late, and another plasmid containing the tissue-specific promoter I am interested in studying. Of course, it is essential that both transcripts be easily distinguishable by separation of the products by electrophoresis. The easiest way to measure transcripts is by using the G-free cassette developed by Sawadogo & Roeder (1985). The use of this template is described in detail in the following section, although other methods can also be used (see section III.B.3.4). As explained in a later chapter, the use of the classical run-off assay is discouraged, since extracts prepared as described contain a variety of enzymatic activities that are incompatible with this assay (see section on "Run-off. Fate of Template"). Similarly, when first testing an extract, it is safer to do an assay in which transcripts can be electrophoretically distinguished from background. In general, background is rather low, and most incorporated counts are derived from correctly initiated transcripts. Occasionally this is not the case, however, and a rapid filter assay could be potentially misleading.

When the G-free cassette or another comparable assay is used, testing the extracts by *in vitro* transcription takes one day. With a little practice, it is possible to do the *in vitro* transcriptions and a set of footprints at the same time. Setting up the reaction takes about half an hour, incubations are for a total of 55 minutes, and processing the sample takes slightly over one hour. Then the gel can be loaded, run, fixed and dried in about 3 hours. A good extract should give a clear autoradiographic signal after 1 to 2 hours of exposure to Kodak XAR-5 film, when using a good quality intensifying screen.

The G-free Cassette

In 1985, Sawadogo & Roeder published a sensitive and rapid assay for measuring *in vitro* transcripts, based on the so-called G-free cassette (or C_2AT cassette). Briefly, this assay relies on the utilization of a synthetic DNA template of 394 bp, synthesized randomly, but using only three nucleotides in the coding strand (C, A and T, hence its name C_2AT). After cloning this DNA in a double -stranded vector, it can be used as a template for transcription *in vitro*, provided that an active promoter is placed in front of the cassette. The advantage of this system is that, when used in the correct orientation, the cassette does not contain G residues in its "coding" strand; therefore, *in vitro* transcriptions can be performed in the absence of GTP. As a result, all incorporated counts are derived solely from transcription from the G-free cassette, since spurious initiation sites elsewhere (mainly in the vector) give rise to transcripts that can not be elongated past the first G in the sequence. Therefore, without further manipulations, total transcription reactions can be displayed on acrylamide gels. The main radioactive band observed is derived from transcripts that are correctly initiated and terminate at the first G residue after the G-free cassette, usually a GGG trinucleotide coming from the Sma I site of the pUC polylinker. Appropriate controls have shown that no transcripts are observed using a promoterless version of this vector in conjunction with nuclear extracts prepared as described (Figure 8).

The only limitation in the use of this elegant method lies in the necessity to clone the promoter of interest in front of the G-free cassette in such a way that no G residues are introduced between the cap site and the G-free cassette. In some cases, luck has provided that the first G lies within a restriction site, but sometimes the cap site can be a G, followed by a G-rich stretch. This obviously complicates matters, specially if these nucleotides happen to be important for transcription! In any case, the extra effort maybe required in cloning is absolutely compensated by the simplicity of the assay.

It is a common temptation to try to get cleaner results by doing the *in vitro* transcription and/or treating the products of the reaction with RNase T_1, which cleaves after G residues. This should be absolutely avoided, since such a treatment masks non-specifically initiated transcripts that transverse through the G-free cassette. By doing this, transcripts incorrectly initiated upstream of the cap site are falsely made to appear as correctly initiated ones.

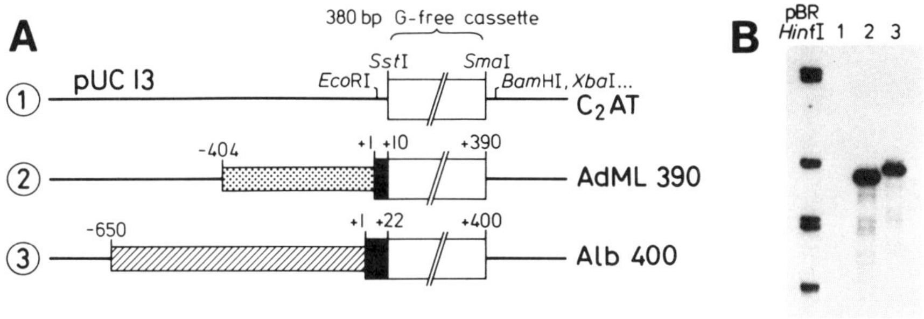

Figure 8.
Transcripts Synthesized *In Vitro* from the G-free Cassette under the Control of Two Different Promoters. A. Diagram of the G-free cassette vector (C2AT) and the two promoter-containing constructs. The 5'-flanking region of each gene is shown either stippled (Adeno Major Late) or hatched (mouse albumin), and the sequences downstream of the cap site are represented in black. B. *In vitro* transcription products. The three templates just described were assayed in a standard *in vitro* transcription reaction using liver nuclear extracts, and the products were displayed on a 4% polyacrylamide/urea gel. Lane 1: Promoter-less cassette vector, Lane 2: G-free cassette under the control of the Adeno Major Late promoter, Lane 3: G-free cassette under the control of the mouse albumin promoter (Gorski et al., 1986).

3.1 Template Preparation

When using the nuclear extracts described in this book, and with the templates we have tested, there is no significant difference using supercoiled or linear templates. In fact, we have found that, when mixed with nuclear extracts, both linear and supercoiled templates become rapidly transformed into high molecular weight species, probably through processes involving ligation and/or concatenation.

Also, in our experience, ultra-clean, twice-CsCl purified DNA is not necessary, and sometimes it is even not as good as fast, "dirty" preparations, probably due to incompletely removed ethidium bromide. There are many rapid methods for preparing reasonably clean DNA; here, I give the protocol I use routinely to prepare DNA to be used as template for *in vitro* transcription reactions (in fact, I use the same DNA preparations for all other purposes as well).

Materials and Buffers

- 37°C shaker incubator
- 250-ml flasks with cotton plugs
- High speed centrifuge
- Swing out rotor (50-ml tube capacity)
- 37°C water bath
- Plastic 10-ml pipets
- Siliconized glass wool
- Mini gel and power supply

Chemicals:

- L-broth or other liquid media for bacterial growth
- 10 mM Tris · HCl, pH 8.0 / 1 mM EDTA (called 10/1 in the protocol)
- 0.2 M NaOH / 1% SDS / 10 mM EDTA, freshly made
- "5 M" Potassium Acetate: To 60 ml of 5 M KAc, add 11.5 ml of glacial acetic acid and 28.5 ml of H_2O. The resulting solution is 3 M with respect to potassium, and 5 M with respect to acetate (from Maniatis et al., 1982).
- 5 M LiCl
- Isopropanol
- 10 mM Tris · HCl, pH 8.0 / 2 mM EDTA (called 10/2 in the protocol)
- 10 mg/ml RNase A, heat treated
- RNase T_1
- 10% SDS
- 0.5 M EDTA
- Bio Gel $A_{1.5m}$ (Bio Rad)
- 5 M NaCl

- Phenol / Chloroform
- Ethanol
- Dry ice
- 10 mM Tris · HCl / 0.1 mM EDTA (called 10/0.1 in the protocol)

Step-by-Step Procedure

Template Preparation

> Grow a 50-ml overnight culture of the bacteria containing the required plasmid in a 250-ml beaker, with vigorous agitation. For pUC-based clones, I do not amplify the cultures with chloramphenicol.

> Spin down the bacteria at 8,000 rpm for 10 minutes.

> Discard the supernatant (with due regard to the NIH Guidelines for Recombinant DNA Research) and vortex the pellet extensively.

> Resuspend the bacteria in 2.5 ml of 10/1. Make sure all bacteria are resuspended, then keep at room temperature for 5 minutes, with occasional agitation.

> Lyse the cells by addition of 5 ml of the freshly made NaOH/SDS/EDTA solution.

> Leave again for 5 minutes at room temperature. Mixing at this step should be done very gently.

> Add 750 µl of "5 M KAc". Leave on ice for 5 minutes, with occasional gentle swirling.

> Spin at 9,000 rpm for 5 minutes in the cold.

> Transfer the supernatant to a new tube containing 11.25 ml of cold 5 M LiCl. Invert to mix and incubate for 60 minutes at –20°C.

> Spin at 9,000 rpm for 5 minutes in the cold.

> Transfer the supernatant to a new tube containing 10 ml of isopropanol. Mix well and leave 5 minutes at room temperature.

> Spin at 9,000 rpm for 15 minutes, at room

temperature or in the cold. Discard the supernatant.

> Dissolve the pellet in 0.5 ml of 10/2.

> Add RNase A to 100 μg/ml and RNase T_1 to 100 Units/ml. Incubate at 37°C for 30 minutes.

> Add 50 μl of 10% SDS and 100 μl of 0.5 M EDTA, and load on top of a Bio Gel $A_{1.5m}$ column. This column is made in a plastic 10-ml pipet with the top broken off: pack a small ball of siliconized glass wool at the bottom, then fill with resin until a volume of approximately 12.5 ml, and wash extensively with 10/0.1.

> Collect approximately 4 ml and discard. Then, collect 10 × 0.5 ml fractions and analyze a 2 μl aliquot of each on a mini gel (0.7% agarose).

> Pool the fractions that contain DNA and add SDS to 0.2% and NaCl to 0.5 M.

> Extract 3 times with phenol/chloroform.

> Precipitate with 2.5 volumes of ethanol. Leave in dry ice for 20 minutes.

> Spin and resuspend in 360 μl of 10/0.1. Transfer to Eppendorf tube.

> Measure the A_{260} and calculate the concentration of DNA.

> If necessary (too diluted), add 40 μl of 3 M NH_4Ac and 1 ml of ethanol. After dry ice and centrifugation, resuspend at 1 mg/ml in 10/0.1.

3.2 Transcription Reaction

There are essentially two variations to this protocol: the transcripts can be made radioactive for direct visualization by gel electrophoresis, or they can be made cold, and then visualized by S_1 analysis or other techniques. The standard protocol uses radioactive UTP and contains O-Methyl GTP instead of GTP. This is a GTP analogue that, when incorporated into a growing RNA chain, prevents further elongation. This protocol was specifically designed for transcription reactions using the G-free cassette described in the introduction to section III.B.3. This is the protocol first described in this section. Two addenda are given at the end of the protocol. The first contains modifications that should be incorporated when templates other than the G-free cassette are used. The second addendum briefly describes the preparation of acrylamide/urea gels for analysis of the results.

Materials and Buffers

- > Water baths (30°C and 37°C)
- > Cold Eppendorf-type centrifuge
- > Regular ice and dry ice
- > Nuclear extract and DNA templates
- > [α-^{32}P] UTP, ~400 Ci/mMole, 10 mCi/ml
- > RNasin or other RNase inhibitor (RNasin, from Promega Biotec, comes at approximately 50 Units/µl)
- > Nuclear dialysis buffer (See Protocol 2, Extract preparation)
- > 10 mg/ml tRNA (commercial tRNA should be phenol extracted and ethanol precipitated before use)
- > 10 mg/ml Proteinase K, dissolved in H_2O and stored in 100-µl aliquots at –20°C
- > Phenol/chloroform (1:1)
- > Ethanol
- > Formamide dye (0.5% Bromophenol Blue, 0.5% Xylene Cyanole FF, 1 mM EDTA in 95% recrystallized formamide)
- > Pre-cast and pre-electrophoresed 4% polyacrylamide/urea gel (see *Addendum 2*)
- > High power electrophoresis power supply
- > Large tray for fixing the gel
- > Gel drier
- > Whatmann 3 MM paper
- > X-ray film and developing facility

Buffers:

5 × MIX:

For 2 ml of mix:

100 µl 1 M HEPES, pH 7.6
350 µl 87% glycerol
85 µl 3 M KCl
60 µl 1 M $MgCl_2$
250 µl 25 mM CTP
250 µl 25 mM ATP
35 µl 10 mM UTP
200 µl 10 mM O-Methyl GTP
670 µl H_2O

This solution should be stored frozen at –20°C, and it can be thawed and refrozen many times.

Transcription Stop Buffer:

	For 100 ml
250 mM NaCl	5 ml, 5 M
1% SDS	10 ml, 10%
20 mM Tris·HCl, pH 7.5	2 ml, 1 M
5 mM EDTA	1 ml, 0.5 M

This solution should be stored at room temperature.

Step-by-Step Procedure

In Vitro Transcription (with G-free Cassette)

> Unless specified, all manipulations are carried out on ice.

> The total volume of the reaction is 20 μl.

> Prepare the following mix (per sample):

4 μl 5 × mix
0.7 μl [α-^{32}P] UTP
1 μl RNasin (Promega Biotec, 50 Units)
1.3 μl H_2O

> Set up a series of three reactions for each extract to be tested, so that the percentage of extract in the reaction can be varied between 30 and 60%:

	Nuclear Extract	Dialysis Buffer
30%	6 μl	6 μl
45%	9 μl	3 μl
60%	12 μl	–

> To each tube, add 1 μl template DNA (this can be a mixture of two or more different templates, but the total amount of DNA should be kept constant at 0.8–1 μg per reaction).

> Incubate on ice for 10 minutes.

> Add 7 μl of the mixture described just above.

> Incubate at 30°C for 45 minutes.

> Add a mix of: 274 μl Transcription Stop Buffer
2 μl 10 mg/ml tRNA
4 μl 10 mg/ml Proteinase K.

> Incubate at 37°C for 30 minutes.

- Extract once with phenol/chloroform.
- Add 750 µl of ethanol (no more salt required).
- Precipitate in dry ice for 15–20 minutes.
- Spin 10 minutes in a cold Eppendorf-type centrifuge.
- Discard the supernatant (it is very radioactive!) and resuspend the pellet, without drying it, directly in 5 µl formamide dye.
- Incubate for 2 minutes in boiling water (I use a microwave oven to boil the water, but do not put the samples in the microwave oven).
- Load 3 µl on a pre-electrophoresed 4% polyacrylamide/urea gel (see *Addendum 2*).
- Run the electrophoresis at a constant power such that the temperature of the gel is approximately 55°C, until the Bromophenol Blue (BPB) reaches the bottom of the 30-cm long gel (in the apparatus I use, this means 38 Watts for 90 minutes).
- Remove the glass plates with the gel from the electrophoresis apparatus and carefully separate the glass plates, leaving the gel attached to one of them (usually, but not always, the one that was not siliconized).
- With a hand-held Geiger counter, localize the position of the unincorporated UTP (strong band just above the BPB). Using the edge of the glass plate not adhered to the gel, cut this portion out. This step mainly serves to reduce contamination of the fixing solution and gel drier.
- Put the gel and the glass plate into a tray containing 20% ethanol/10% acetic acid. Fix for 15–20 minutes.
- Remove the glass plate and gel from the fixer, carefully drain by letting it stand almost vertically for a few seconds, and then transfer the gel to a dry piece of Whatmann 3 MM paper.
- Dry under vacuum (between 15 minutes and 2

hours, depending on the drier and the strength of the vacuum).

> Expose to X-ray film for 1 to 2 hours (I use Kodak XAR-5 film and Kodak Lightning Plus intensifying screens).

Addendum 1: Modifications Required when not using the G-free Cassette

If the transcripts are to be measured by primer extension, RNase mapping, S_1, or any other similar technique, the following modifications must be made:

- The 5 × mix described should be replaced by:

  5 × Cold Mix:

  100 µl 1 M HEPES, pH 7.6
  350 µl 87% glycerol
  85 µl 3 M KCl
  60 µl 1 M $MgCl_2$
  250 µl 25 mM GTP
  250 µl 25 mM ATP
  250 µl 25 mM UTP
  250 µl 25 mM CTP
  405 µl H_2O

- The reaction is carried out exactly as before, except that radioactive UTP is replaced by the same volume of H_2O. I routinely double the size of the reaction when not using the G-free cassette.
- After the reaction is finished, it is important to remove the template DNA, which otherwise might also hybridize with the probes used. For this purpose, the following protocol has been used:

After Proteinase K digestion, phenol/chloroform extraction and ethanol precipitation as usual, resuspend the pellet in 10 µl of H_2O, and then add 90 µl of:

	For 10 ml
20 mM HEPES, pH 7.6	200 µl, 1 M
50 mM NaCl	100 µl, 5 M
10 mM DTT	100 µl, 1 M
5 mM $CaCl_2$	50 µl, 1 M
5 mM $MgCl_2$	50 µl, 1 M

containing 3–5 µg RNase-free DNase I and 1 µl RNasin per sample.

The solution (without DNase and RNasin) is stored at –20°C and is stable upon repeated freezing and thawing, but DTT is slowly degraded, and sometimes has to be re-added. For RNase-free DNase, I routinely use iodoacetate-treated enzyme.

– Incubate at 37°C for 15 minutes.

– Add 100 µl of 2 × Stop Buffer:
  2 ×: 20 mM Tris · HCl, pH 8.0
  50 mM EDTA
  1% SDS
  0.3 M NaCl

– Heat at 65°C for 5 minutes.

– Extract twice with phenol/chloroform.

– Precipitate with 500 µl of ethanol (no more salt required).

– Wash RNA pellet in 70% ethanol and dissolve in H_2O without drying the pellet. This RNA is ready for further analysis.

Addendum 2: Preparation of Polyacrylamide/Urea Gels

Since these gels are routinely used for many analyses in molecular biology (sequencing and other methods), I will not give a detailed description on how to prepare the gels. I will instead emphasize the details that might be relevant to their particular application in the analysis of *in vitro* transcripts. The gels used for this purpose are in no way different from regular sequencing gels.

When using the G-free cassette, the transcripts produced are about 400 nucleotides in length (smaller versions of the cassette are also routinely used). Since this is already close to the upper limit of accuracy for polyacrylamide gels, I use only 4% acrylamide. These gels are sufficiently sturdy to be handled easily, but low enough in acrylamide concentration

as to allow good resolution. Obviously, when methods other than the G-free cassette are used to analyze the transcripts, the concentration of acrylamide should be varied according to the length of the expected band. The gels I use have the usual thickness for sequencing gels (0.2–0.4 mm) and contain the usual concentration of urea (8.3 M). They are 30–40 cm in length and 15 cm in width. I do not de-gas the acrylamide before pouring the gel, nor do I take special precautions to avoid ribonucleases (the urea and high temperature should take care of them). I do routinely pre-electrophorese the gels for about 30 minutes, but have not noticed any difference when this step was "accidentally" omitted.
In some cases, results have been analyzed by electrophoresis through glyoxal gels. Due to the lower resolving power of agarose, these gels do not look as pretty as acrylamide gels.

3.3 Analysis of Results

Four parameters should be evaluated when an extract is tested for the first time in an *in vitro* transcription assay: strength of the signal, background level, dependence on protein concentration, and tissue specificity.

Strength of the signal: When using liver, spleen or brain extracts (these are the ones I have used most frequently, and can therefore generalize), reasonably strong signals are obtained from the G-free cassette under control of the Adeno Major Late promoter after 1 or 2 hours of exposure. A signal that requires overnight exposure denotes a weak, but probably still usable extract; however, if only a weak signal is observed overnight, then I consider the extract as dead. This could be due to several different reasons, some of which are discussed in the trouble-shooting section. Obviously, different strengths will be observed when using different promoters. Similarly, if other assays, such as S_1, are used, the strength of the signal will depend on the specific activity of the probe. Therefore, only practice can help the evaluation of results with respect to the efficiency.

Dependence on protein concentration: Usually, the strength of the signal changes with the amount of extract used in the reaction. This change is not the same for all promoters, but for several liver-specific genes, it has been found that optimal transcription is achieved when the extract represents 45% of the total volume in the reaction (since most of the time the extracts have a protein concentration between 7 and 9 mg/ml, this represents a final protein concentration of 3 to 4 mg/ml). Occasionally, when the extracts are too concentrated, a lower percentage of extract in the reaction has to be tested in order to find the optimal conditions. Unfortunately, it is not possible to test concentrations of more than 60% (12 μl)

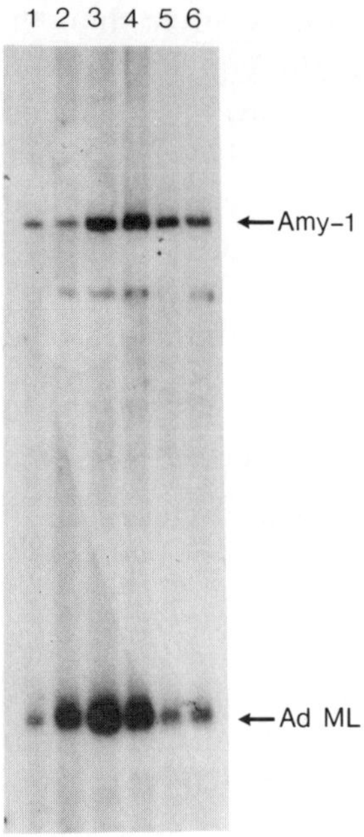

Figure 9.
Effect of Protein Concentration on the *in vitro* Transcription Reaction. The concentration of liver nuclear extract was varied as follows: Lane 1: 0.7 mg/ml, Lane 2: 1.4 mg/ml, Lane 3: 2.1 mg/ml, Lane 4: 2.8 mg/ml, Lane 5: 3.5 mg/ml, Lane 6: 4.2 mg/ml. In all cases, the template was a mixture containing 100 ng of Adeno Major Late promoter in front of a shortened version of the G-free cassette, plus 900 ng of the parotid-specific Amy-1 promoter in front of the regular length G-free cassette.

of extract, so that very diluted extracts (less than 5 mg/ml of protein) are rarely active in transcription assays.

Tissue specificity: When first testing an extract, it is advisable to use as template a mixture consisting of a plasmid containing the G-free cassette under the control of the tissue-specific gene of interest, and a plasmid containing a shorter version of the same cassette, under the control of a promiscuous promoter, such as the Adeno Major Late (AdML). It seems that this viral promoter requires few transcription factors for its activity, when compared to most tissue-specific promoters. Therefore, analysis of transcriptional activity based solely on the activity of the Adeno promoter can be potentially misleading. To properly assess tissue specificity, the two extracts to be compared should utilize the AdML promoter with similar efficiency, while transcription from the tissue-specific promoter in question should be greatly diminished in the extract from the non-expressing tissue. For the most part, however, tissue-specific promoters are not completely silent in extracts from non-expressing tissues; rather, a low, basal level of activity is always observed.

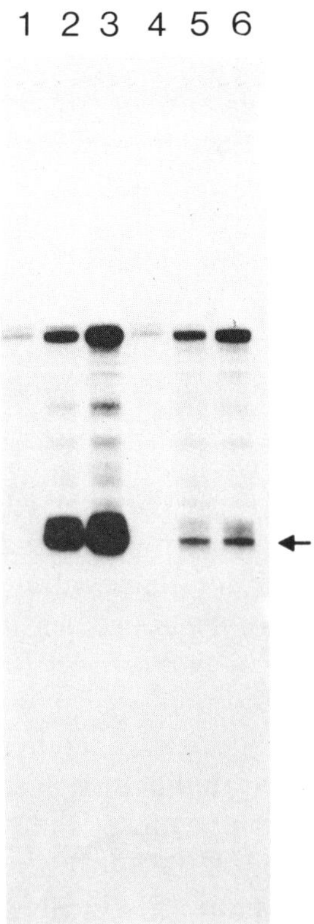

Figure 10.
Tissue Specificity of the *in vitro* Transcription Reaction. *In vitro* transcription reactions were performed with the following templates: Lanes 1 and 4: pUC 19, Lanes 2 and 5: α2U-globulin promoter (from –3000 to +65), Lanes 3 and 6: α2U-globulin promoter (from –376 to +65). Each template was transcribed in nuclear extracts from either liver (Lanes 1 through 3) or spleen (Lanes 4 through 6). The arrow indicates correctly initiated transcripts, as determined by an S_1 assay (Sierra et al., 1990).

3.4 Alternative Assays

In some cases, the use of the G-free cassette is not possible. This could be due simply to technically cumbersome cloning procedures (for example, if no convenient restriction sites exist close downstream of the promoter, or the sequence between the start site and existing restriction sites is G-rich). On the other hand, there are cases in which sequences immediately downstream of the cap site are important for transcription, and cannot be deleted without impairing the function of the promoter (Smale & Baltimore, 1989). Finally, many genes contain more than one cap site, and removal of some of these sites can have unexpected effects. In all of those cases, it might be unwise to use the G-free cassette assay just described. Many of the assays currently used to detect *in vivo* or *in vitro* correctly initiated transcripts can be used to assay *in vitro* transcripts produced using the nuclear extracts described in this book. However, some precautions must be observed in order to obtain reliable answers. Some of these assays have been tried (with varying success) in different laboratories, and I present here a list of the methods and their potential pitfalls.

3.4.1 S_1 analysis

This method has been very successful for the detection of *in vitro* synthesized transcripts. By using regular CIP/kinasing protocols, probes can be made hot enough that the time required for the detection of protected bands after gel electrophoresis is quite comparable to that required to observe transcripts directly from the G-free cassette (1 to 2 hours), when testing a strong liver-specific promoter, such as those from albumin or α2U-globulin.

The main precaution that should be kept in mind when measuring in vitro transcripts by S_1 analysis is that the nuclear extracts usually contain significant amounts of endogenous hnRNA. These will hybridize with the probe, thus giving a positive signal even when no transcription has occurred *in vitro*. To counteract this potential problem, care must be exerted to choose the probe in such a way that the label is located within vector sequences present in the clone to be transcribed, rather than within eukaryotic sequences derived from the gene unit under analysis. Protected bands of the right size can then be safely assumed to correspond to transcripts synthesized *in vitro*.

Full-length probe protection is routinely observed when analyzing *in vitro* transcripts by S_1, even when the probes are strand separated before use. When strong promoters are used, enough of the transcripts appear to be derived from polymerases transversing the whole plasmid, thereby transcribing, at the end of a round of transcription, promoter sequences located upstream from the initiation site. α-amanitin studies have shown, however, that some of the full length protection of the probe is actually not due to transcription by pol II. No exhaustive analysis has been attempted, though, so it is not clear at present whether these transcripts are the result of pol I or pol III activities, or both.

Step-by-Step Procedure

Probe preparation

The protocol I use is exactly as described by Maniatis et al. (1982). Whenever possible, single-strand probes are prepared by running the labelled probe on a strand separation gel as follows:

> Pour a preparative (1.5 mm thick) acrylamide gel in TBE buffer, without urea and containing a single large well for sample loading. For these gels, a special mix containing a 50:1 ratio of acrylamide to bis-acrylamide is used.
> For selecting the acrylamide concentration, use the following rule of thumb (adapted from Maniatis et al. (1982)):

Fragment Length (in bp)	% Acrylamide in Gel
900–500	3
500–200	4
80–200	5
40–100	8

Fragments larger than 1 Kb are difficult to strand separate (it depends on base composition as well as specific sequences, i.e., on luck). I do not recommend the use of probes smaller than 50–70 bp, since hybrids below this range can be very unstable, thus giving irreproducible results. While the above rule of thumb gives an idea on where to start, it is advisable to run some analytical gels before putting the whole sample on a gel.

> Set up the gel in the electrophoresis apparatus, do not pre-electrophorese.

> In the meantime, denature the sample as follows:
> 30 µl CIP/kinased DNA, in H_2O
> 30 µl 0.2 N NaOH (freshly diluted)

let stand at room temperature for 10 minutes, then heat to 50°C for 10 minutes

- To the denatured sample, add 30 µl of Bromophenol Blue/ficoll mix (preheated to 50°C) and load the whole sample as fast as possible on the gel.
- Run at 350 Volts until the dye has migrated about 2 cm (approximately 30 minutes), then lower the voltage to 200 Volts or less for the rest of the run (usually overnight or longer).
- After electrophoresis, separate the plates, and expose the wet gel to an X-ray film for 10 minutes (or up to a couple of hours if the probe is not very hot). With a sharp scalpel, cut out the band of interest and electroelute (many electroelution apparatus are available; follow the instructions from the manufacturer). If no electroelution apparatus is available, use the protocol described in Maniatis et al. (1982).

Note:
In order to determine which of the strands should be used for S_1 experiments, I do a trial labelling with Klenow enzyme. This is easier than CIP/kinase, so I can do a fast experiment in which only the correct strand is labelled (cut first with one enzyme, label, then cut with the second enzyme). By running this sample in parallel with the CIP/kinased sample, (which is labelled at both ends, thus avoiding lengthy treatments after labelling), it is possible to determine which of the strands is the good one (it co-migrates with the Klenowed sample).

Hybridization and S_1 analysis

Buffers:

10 × Hybridization Salts:

0.4 M PIPES, pH 6.4
10 mM EDTA
4 M NaCl

Store at –20°C.

10 × S_1 BUFFER:

2.8 M NaCl
0.3 M Na Acetate, pH 4.4
45 mM Zn Acetate

Stop Mix:

2.5 M NH_4 Acetate
50 mM EDTA

- Dissolve the RNA sample (from *Addendum 1* to protocol 3.2) in 2 µl H_2O + 2 µl single-stranded probe (at 125,000 cpm/µl).
- Add 32 µl of recrystallized formamide. Leave at room temperature for about one hour, with occasional mixing. This is to ensure that all the RNA is well resuspended.
- Add 4 µl 10 × hybridization salts.
- Incubate at 50°C overnight.
- The next day, very quickly spin two tubes at a time, for as short a time as possible in an Eppendorf-type centrifuge at room temperature. Return the samples to the water bath. This must be done as fast as possible to prevent changes in the hybridization due to the change in temperature.
- Mix the following (per sample):
  267 µl H_2O
  30 µl 10 × S_1 buffer
  1.85 µl 4 mg/ml salmon sperm DNA
  1 µl S_1 nuclease (~100 Units)
- Add 300 µl of this mix to each sample, while the tube is still at 50°C.

Immediately remove the tube from the water bath, mix well, and transfer to a water bath at 37°C.

- Incubate at 37°C for 30 minutes.
- Add 75 µl of a mix containing (per sample):
  73 µl Stop mix
  2 µl 10 mg/ml tRNA
- Add 400 µl of isopropanol. Incubate at –20°C for 10 minutes
- Spin 10 minutes in a cold Eppendorf-type centrifuge. Discard the supernatant (radioactive) and dissolve the pellet, without drying, in a buffer suitable for gel electrophoresis analysis.

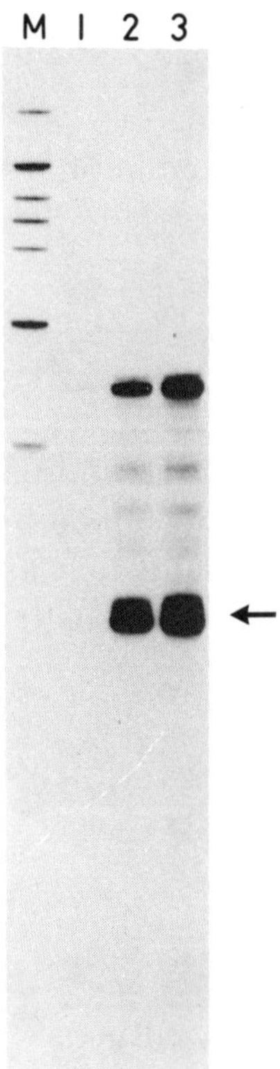

Figure 11.
Analysis of *in vitro* Transcripts by S_1. *In vitro* transcription reactions were performed with the following templates: Lane 1: pUC 19, Lane 2: α2U-globulin promoter (from –3000 to +65), Lane 3: α2U-globulin promoter (from –376 to +65). The single-stranded S_1 probe spanned from position –74 to the Hind III site in the pUC polylinker. The arrow indicates the position of correctly initiated α2U-globulin *in vitro* transcripts (102 nucleotides), and the higher molecular weight band present in all lanes corresponds to the entire length of the S_1 probe (see text for discussion). (Sierra et al., 1990).

3.4.2 RNase Mapping

This method has also been successfully used to analyze *in vitro* synthesized transcripts. In this case, however, the potential for problems with hnRNA derived from the nuclear extracts is compounded by the fact that the probe is internally labelled. This problem can again be avoided by using a riboprobe that starts far into vector sequences. This way, endogenous RNA gives a protected fragment considerably shorter than *in vitro* synthesized RNA, which contains the same vector sequences as the probe. Nevertheless, results could be difficult to interpret with this technique if the gene in question contains several start sites, such that a battery of bands would be obtained, due to the presence of both *in vivo* and *in vitro* transcripts.
Essentially, I have used the basic techniques for probe preparation and RNase mapping as described by Melton et al. (1984). Of course, polymerases other than SP6 can be used, provided the appropriate vector is used to clone the promoter fragment to be used as a probe.

Step-by-Step Procedure

RNase Mapping

Probe preparation

Buffers:

10 × SP_6 Buffer:

0.4 M Tris · HCl, pH 7.5
60 mM $MgCl_2$
0.1 M DTT

> Mix the following at room temperature:

1 µl 10 × SP_6 buffer
0.5 µl NTP (-UTP), 10 mM each
0.5 µl 2 mM UTP
1 µl 1 mg/ml BSA
1 µl DNA template (~1 µg, truncated)
4 µl [α-^{32}P] UTP, (~400 Ci/mMole, 10 mCi/ml)
1 µl RNasin (Promega Biotec, 50 Units)

> Let stand at room temperature for 2 minutes.

> Add: 1 µl 20 mM spermidine
> 0.5 µl SP_6 polymerase (10 Units)

> Incubate at 40°C for 60 minutes.

> Add 5 µl 10 mg/ml tRNA and 1 µl (~750 ng) of iodoacetate-treated DNase I.

Note:
This is much more DNase I than suggested by Melton et al., but I have not had problems with contaminating RNases.

> Incubate at 37°C for 10 minutes.

> Add 3 µl of 0.5 M EDTA and 70 µl of oligo (dT) application buffer.

> Extract once with phenol/chloroform.

> Run the aqueous phase directly through a small Bio Gel $A_{1.5m}$ column, prepared in a Pasteur pipet, and equilibrated with 10 mM Tris · HCl, pH 8.0/0.1 mM EDTA.

> Collect the flow through (void volume) and count a small aliquot to determine the degree of incorporation. Usually, this can be used directly to do the hybridization, since it should contain more than 250,000 cpm/μl.

Hybridization and RNase mapping

Buffers:

10 × RNase Mapping Buffer:
3 M NaCl
0.1 M Tris · HCl, pH 7.5
50 mM EDTA

Store at –20°C.

The hybridization is done exactly as described for S_1 analysis, except that the probe should be at 250,000 cpm/μl. Hybridization temperature is maintained at 50°C.

> After overnight hybridization, put samples on ice.

> Add 300 μl per sample of the following mix:

268 μl H_2O
30 μl 10× RNase Mapping Buffer
1.2 μl 10 mg/ml RNase A
0.6 μl RNase T1 (60 Units)

> Incubate at 30°C for 60 minutes.

> Add 20 μl 10% SDS. Then, add 10 μl 10 mg/ml proteinase K.

> Incubate at 37°C for 30 minutes.

> Extract twice with phenol/chloroform.

> Add 2 μg (not more) of tRNA and ethanol precipitate with 750 μl of ethanol (no more salt required).

> Resuspend in a buffer suitable for gel electrophoresis analysis.

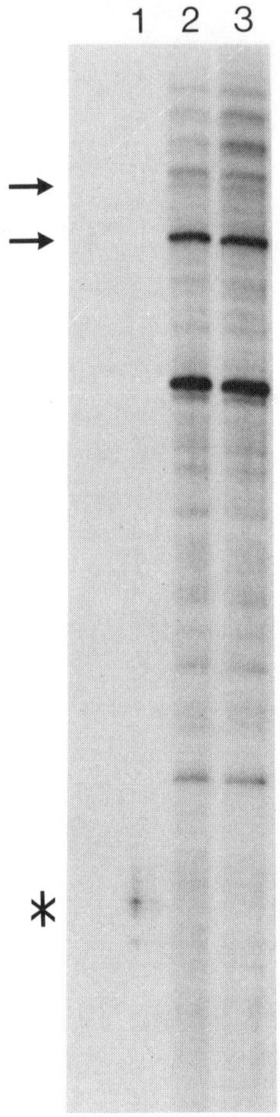

Figure 12.
Analysis of *in vitro* Transcription Reactions by RNase Mapping. *In vitro* transcription reactions were performed with the following templates: Lane 1: pUC 19, Lanes 2 and 3: Major Acute Phase protein promoter in front of the CAT gene. The riboprobe utilized spanned from –106 to the EcoRI site within the CAT gene. The arrows indicate the position of correctly initiated Major Acute Phase protein *in vitro* transcripts (from the two major cap sites of the gene), while the asterisk indicates the position of endogenous transcripts initiated at the upstream-most major cap site. Notice that, while correctly initiated transcripts are easily detectable, there is a considerable number of non-specific bands, not usually seen when *in vivo* transcripts are analyzed by this technique.

3.4.3 Primer Extension

In principle, this method should be perfectly suitable for analysis of *in vitro* transcripts. However, in my hands, primer extension is considerably less sensitive than S_1 or RNase mapping performed as previously described. The method has been successfully used to detect *in vitro* transcripts (see for example, Freedman et al., 1989). Nevertheless, I will refrain in here from providing a detailed technique that has not been satisfactory in my hands.

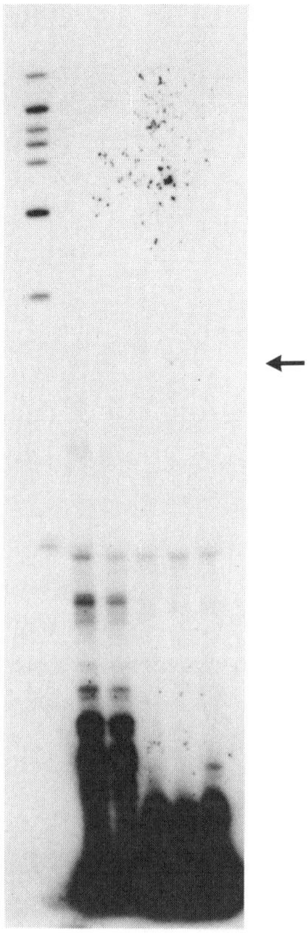

Figure 13.
Analysis of *in vitro* Transcription Reactions by Primer Extension. A construct containing the α2U-globulin promoter was transcribed *in vitro* using liver nuclear extracts, and the products were analyzed by primer extension using five different techniques available in the literature. The primer (a 31-mer) was complementary to sequences present in the vector, and, if correctly extended to the expected cap site, a 136-nucleotide-long band would have been obtained (arrow).

3.4.4 Run-off. Fate of Template

Often, run-off is the method of choice for the analysis of *in vitro* transcripts. The assay is simple, fast, and usually trouble- free. However, many activities present in the extracts prepared as described in this book prevent this assay from being of general use. Early observations in U. Schibler's laboratory indicated that the efficiency of *in vitro* transcription from a variety of templates containing the G-free cassette was similar whether a linear or a supercoiled template was used. Further investigations indicated that, when linear templates not containing the G-free cassette were used, transcription occurred *in vitro* at the usual rate, however, no run-off transcripts of the expected size were observed when the products were analyzed by gel electrophoresis. Instead, very large transcripts were produced. This prompted the experiment shown in Figure 14, in which the fate of the DNA template during the course of an *in vitro* transcription reaction was followed.

The experiment with circular templates shows that, very early during the reaction, activities present in the extract are apparently capable of concatenating the template plasmid, which is then no longer able to enter the gel. Studies with linearized plasmids have revealed an array of active DNA processing activities present in the nuclear extracts. As a result of these activities, linear templates become ligated into larger structures.
In trying to avoid these activities, I prepared templates with non-cohesive ends, as well as templates that had been dephosphorylated at the 5' ends by alkaline phosphatase. Neither of these treatments prevented the subsequent repair by the nuclear extracts, as evidenced again by the synthesis of transcripts that were much longer than expected from run-off transcription, as well as by direct visualization of the template DNA as shown in Figure 14.

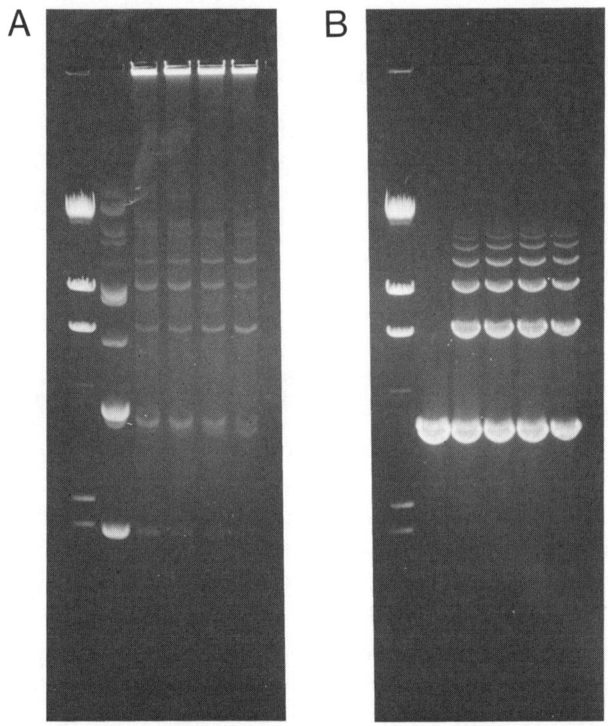

Figure 14.
Fate of Template DNA During the Reaction. *In vitro* transcription reactions were performed as usual, using either a circular (Panel A) or a linear (panel B) template. Aliquots were removed at 0, 5, 10, 20 and 45 minutes, digested with proteinase K as usual, and nucleic acids were extracted with phenol/chloroform. The structure of the DNA template was analyzed in each case by native agarose gel electrophoresis in the presence of ethidium bromide.

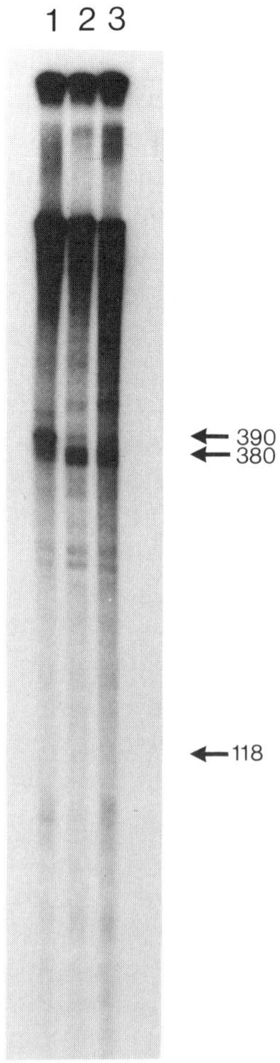

Figure 15.
Analysis of *in vitro* Transcription Reactions by a Run-off Assay. *In vitro* transcription reactions were performed with truncated templates in the presence of [α-^{32}P] UTP, and the products were analyzed on a 4% polyacrylamide gel. The templates were: Lane 1: Adeno Major Late promoter (expected run-off transcript of 390 nucleotides), Lane 2: α2U-globulin promoter (expected run-off transcript of 118 nucleotides), Lane 3: α2U-globulin promoter (expected run-off transcript of 380 nucleotides).

I should nevertheless emphasize that, even though I have not been able to reproducibly synthesize run-off transcripts using these nuclear extracts, the total amount of radioactivity incorporated into RNA during these experiments was still proportional to the strength of the promoter used and, by appropriately subtracting the background obtained from a promoterless template, the relative strength of the promoter tested could be roughly estimated. Of course, no reputable journal should accept such crude data, but as a starting point before deciding how much effort to invest in a new project, the approach is more or less adequate.

4 Footprints: Introduction

The measurement of DNA-protein interactions that might be important for transcription is certainly as relevant as transcription itself in assessing results obtained with nuclear extracts *in vitro*. There are several methods to measure these interactions, and the most widely used are footprinting and gel retardation assays. While gel retardations can be used in a quantitative fashion, footprints are less adapted to quantification, although some reports exist on methodology to quantitate footprints. Nevertheless, as a qualitative tool, footprinting is a very powerful technique.

Essentially, with both of these techniques, DNA-protein interactions are allowed to occur *in vitro*, and the results are then analyzed by gel electrophoresis. In the case of gel retardation assays, the DNA must be rather short (no more than 50 bp) and it is used in excess over the proteins to be measured (hence its quantitative nature). In footprinting experiments, the DNA can be longer (up to a few hundred bp) and the footprint is obtained only when the protein to be assayed is in excess (so-called "half-protection assays" are used when the method is meant to be quantitative). After the DNA-protein interaction has been established, it is measured either directly by a shift in mobility imparted to the DNA by its association with the protein (gel retardations) or it can be measured via accessibility of the DNA backbone to specific probes. These probes, which will induce a break in the DNA backbone, can be relatively simple chemicals, such as MPE [Methidium Propyl EDTA-Fe(II)], or enzymatic, such as DNase I. In my experience, MPE works best when the protein to be assayed is already partially purified, but it gives ambiguous results when whole, unfractionated extracts are used. Therefore, I will limit the present discussion to DNase I footprints, as originally described by Galas & Schmitz (1978).

Footprints, as opposed to *in vitro* transcription, do not require the entire transcription machinery to be intact. If a crucial component of this machinery is either partially lost or inactivated during the handling of the extract, *in vitro* transcription will be affected preferentially. The footprints may or may not be immediately affected, since they do not measure function of the protein, only its binding activity. Therefore, a transcription factor could be inactivated during preparation of the extract, such that it will no longer affect transcription. However, the inactivation could be such that its specific interaction with the DNA template is not affected: a footprint will still be present. Conversely, if the DNA binding property is affected, then both the footprinting and the transcriptional effects would be impaired.

Finally, it is widely believed that transcription *in vitro* requires, in addition, the activity of proteins that do not themselves interact with the DNA, but rather, they interact with other proteins (Gester & Roeder, 1988). These effects will be overlooked if only footprinting experiments are performed. The ability of a nuclear extract to perform in an *in vitro* transcription assay therefore validates the data obtained by the more structural approaches, such as footprinting.

The technique of DNase I footprinting follows a logic similar to that of the Maxam and Gilbert chemical method of DNA sequencing. A DNA molecule is labelled only at one of its ends (and only one of the strands). This DNA is then allowed to interact with the nuclear proteins, which will bind to the DNA at specific points, depending on the protein and the DNA sequence. Then, DNase I is added in limiting amounts, such that, on the average, each DNA molecule is cleaved only once. The cutting by DNase I is not random, but will produce a characteristic, reproducible pattern of cleavage on the naked DNA fragment. When this DNA is first allowed to complex with proteins, specific regions of the DNA will be protected by the presence of these proteins, and therefore, will not be available as a substrate for DNase I cleavage. If all the DNA molecules present in the assay have proteins bound to a specific site (i.e., if the particular protein is in excess), no DNase I cleavage will occur within this region. Therefore a white-out space, or footprint, will be observed in that area.

Footprinting is a very simple and rapid assay. Preparation of the probe usually takes almost a full day if you must begin from scratch, but after that, the binding reaction, DNase I digestion and subsequent processing for gel electrophoresis only takes a couple of hours. Then, electrophoresis and processing of the gel takes about 3 hours, and exposure takes between 2 hours and overnight, depending on the specific activity of the probe.

4.1 Probe Preparation

Once preliminary experiments have been done to determine the region of interest, it is convenient to subclone this region in the middle of a polylinker, such as the one from the pUC18/19 plasmids. This avoids the use of costly and/or finicky restriction enzymes each time a new footprinting experiment has to be done.

The DNA fragment to be analyzed (usually between 100 and 800 bp) should be end-labelled by Klenow (or by CIP/kinase) and resuspended in 10 mM Tris · HCl, pH 8.0 /0.1 mM EDTA at a concentration of about 1–3 ng/µl. This should provide between 10,000 and 100,000 cpm/µl.

Materials and Buffers

All equipment required to prepare the DNA probe for footprinting should be standard apparatus and chemicals in any molecular biology laboratory. These include restriction enzymes, their buffers, water baths, Klenow enzyme, radioactive dNTPs, and an electroelution apparatus. If no electroelution apparatus is available, the use of low gelling agarose has been quite satisfactory for the preparation of probes.

Step-by-Step Procedure

Probe Preparation

> Digest the plasmid to be analyzed with a restriction enzyme that cuts only once, near the region of interest. Note that footprints located too close to this site will migrate very fast, and will therefore be located at the bottom of the gel during the final analysis. This step can be done on a large scale, and the digested plasmid can then be stored and used in subsequent labelling experiments.

> Check for complete digestion by running a small aliquot (~100 ng) on a 0.7% agarose mini-gel. Extract the remainder of the DNA once with phenol/chloroform and precipitate with ethanol.

> Resuspend the DNA at a concentration of 1 mg/ml in 10 mM Tris · HCl/0.1 mM EDTA.

> Set up the Klenow reaction as follows:

5 µl digested DNA
2 µl 10 × Hind III buffer
2 µl 40 mM dNTP (-ATP)
6 µl [α-^{32}P]dATP (~3,000 Ci/mMole, 10 mCi/ml)
4 µl H_2O
1 µl Klenow enzyme

Note:
The choice of label depends on the overhang created by the restriction enzyme used. I try, whenever possible, to use a clone in which the label can be put at an Eco RI site, so that up to two radioactive atoms can be incorporated per molecule.

> Incubate at room temperature for 10 minutes (protected by a lead shield).

> Add 2 µl of 4 mM dATP (this is a chasing step: use

an excess of the same nucleotide that was used for labelling).

- Incubate at room temperature for 5 minutes.
- Add 2 µl 10% SDS and 10 µl 0.5 M EDTA. Then extract once with phenol/chloroform and ethanol precipitate.
- Resuspend the radioactive DNA in a small volume of 10 mM Tris · HCl, pH 8.0 /0.1 mM EDTA and digest with the second restriction enzyme. Use an excess (2–3-fold) of enzyme, and add enzyme again after 30 minutes. This way, you avoid having to check for complete digestion, thus avoiding further contamination of a gel apparatus.
- Extract again with phenol/chloroform, and load directly on a preparative agarose gel. Use low gelling temperature agarose if no electroelution is possible.
- Run the gel as long as necessary to obtain good separation between bands. Examine the gel briefly under UV light at 254 nm, and, with a sharp scalpel, excise the band of interest.
- Extract the DNA from the agarose, either by electroelution or by melting the low gelling agarose.
- Resuspend the probe in 50 µl of 10 mM Tris · HCl, pH 8.0/0.1 mM EDTA.
- Count 1 µl in a scintillation counter. It should give around 10,000 cpm.

4.2 Binding Reaction

All components of the binding reaction are kept on ice at all times, including the DNase I digestion step. Subsequently, samples are incubated at higher temperatures. It has been found in our laboratory that results are more reproducible under these conditions than if the experiment is done at room temperature or at 37°C. The time required to form complexes between DNA and nuclear proteins has been measured for several promoters. In general, the reaction is complete after only 2 or 3 minutes of incubation. Just to be safe, and because of practical reasons, I routinely do the binding for 15 minutes, but incubations for up to 90 minutes have shown no obvious difference when compared with shorter binding times. I would guess that this is not absolutely always the case, however, and it is better to keep the binding time rather short.

Gel retardation assays have shown that frequently, Mg^{+2} ions inhibit protein binding to DNA, such that a lower concentration of protein is required to obtain a shift in the absence of these ions. Unfortunately, DNase I requires Mg^{+2} ions for its activity. It is not clear what effects, if any, this phenomena might have on footprinting analysis. In some cases, when the Mg^{+2} ions are perceived to be a potential problem (for example, when the extracts contain large amounts of endogenous DNase I), they can be omitted from both the binding and DNase digestion steps, but in this case, DNase II, rather than DNase I should be used. This enzyme does not require Mg^{+2} for its activity. Unfortunately, it cleaves DNA much less frequently than DNase I. Therefore, results obtained with DNase II are more difficult to interpret than classical DNase I footprinting.

Finally, I would like to mention that, for gel retardation assays, the same binding conditions are used as for footprinting. However, the protein concentration is kept much lower (usually 2–10 µg per reaction, compared to 30–100 µg for footprinting) and the products of the binding reaction are directly loaded on non-denaturing gels.

Materials and Buffers

10 × Footprinting Buffer:

	For 10 ml
250 mM HEPES, pH 7.6	2.5 ml, 1 M
50 mM $MgCl_2$	500 µl, 1 M
340 mM KCl	1.1 ml, 3 M

Nuclear Dialysis Buffer: As described in Protocol 2.

Poly (dI:dC): 2 mg/ml in 150 mM NaCl

DNase I: Stock solution at 5 mg/ml in 2 mM HCl is stored in small (5 µl) aliquots at –70°C.

DNase Dilution Buffer:

	For 10 ml
25 mM $CaCl_2$	250 µl, 1 M
10 mM HEPES, pH 7.6	100 µl, 1 M

Stop Buffer:

	For 100 ml
20 mM Tris · HCl, pH 8.0	2 ml, 1 M
20 mM EDTA	4 ml, 0.5 M
250 mM NaCl	5 ml, 5 M
0.5% SDS	5 ml, 10%

Sonicated Salmon Sperm or *E. Coli* DNA: 4 mg/ml in H_2O

Proteinase K: 10 mg/ml in H_2O.

Step-by-Step Procedure

Footprinting Reaction

> On ice, mix the following:

13 µl of nuclear extract plus Nuclear Dialysis Buffer
2 µl 10 × footprinting buffer
0.5 µl poly (dI:dC)
1 µl probe DNA (1–3 ng)
3.5 µl H_2O

Notes:

- All these components, except for the protein extract, can be pre-mixed at the start of the experiment.
- The protein concentration should not exceed 5 mg/ml (100 µg total), and usually, 30–50 µg of nuclear proteins is sufficient.
- Incubate the above mix on ice for 15 minutes.
- In the meantime, prepare a dilution of DNase I in its dilution buffer, and let it sit on ice for at least 5 minutes. The dilution depends on the amount of protein used:

  1:25 to 1:50 for more than 50 µg/assay
  1:100 for between 10 and 50 µg/assay
  1:200 for 0 to 10 µg/assay

Routinely, at least two of these dilutions are required, since a control in which no proteins are added is always performed. When first testing an extract, several concentrations of protein should be analyzed.

> Add 3 µl of DNase I to each sample. Space them

every 15 seconds, so as to be precise in the time of digestion.

> Incubate on ice for 5 minutes.

> Add 80 µl of a mixture containing, per sample:
78 µl stop solution
1 µl ssDNA or *E. coli* DNA (sonicated)
1 µl proteinase K

> Incubate at 45°C for 60 minutes.

> Extract twice with phenol/chloroform and precipitate with 250 µl of ethanol.

> Resuspend in 5 µl of formamide dye.

> Heat to 90°C for 2 minutes and load 3 µl on a polyacrylamide/urea sequencing gel (the concentration of acrylamide depends on the size of the fragment to be analyzed).

> Run the gel as usual, fix, dry and expose to X-ray film.

4.3 Analysis of Results

The analysis of footprinting results is rather straightforward: white spaces should be observed in specific areas of the probe when proteins are added. If no clear areas are obvious, either the DNA lacks protein binding sites, or the extract is not concentrated enough in nuclear proteins (a high protein concentration is still theoretically possible, if cytoplasmic contamination exists, but this is uncommon). Alternatively, a complete blanking out could occur. This is believed to be due to non-specific binding by proteins such as histones. A more comprehensive list of possible artifacts or anomalous results in footprinting experiments is provided in the trouble-shooting section.

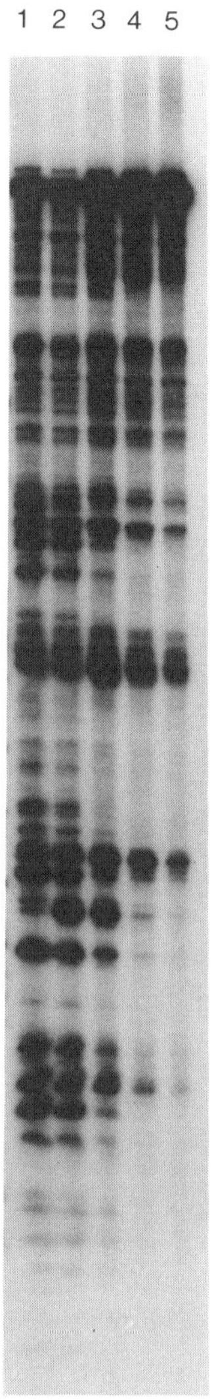

Figure 16.
Effect of Protein Concentration on Footprint Assays. The concentration of male liver nuclear extract required to induce footprints in the promoter region of the α2U-globulin gene was titrated. Lane 1: No nuclear extract added, Lane 2: 175 μg/ml, Lane 3: 0.5 mg/ml, Lane 4: 1 mg/ml, Lane 5: 1.7 mg/ml. In all cases, the concentration of total protein was kept constant by addition of BSA. With this particular combination of extract and promoter, very low concentrations of protein already proved to be saturating, and a careful titration is often desirable when new extracts are being used.

C Factors Affecting *In Vitro* Transcription

Many variables can affect the efficiency and/or specificity of *in vitro* transcription reactions. In this chapter, I will briefly discuss the most important of these variables, whose influence has been methodically analyzed.

The effect of protein concentration has already been discussed, and it has already been emphasized that this variable should be checked each time a new transcription extract is prepared, since the transcriptional ability of an extract depends not only on the total amount of protein in the assay, but also on the type and quality of these proteins. Other important variables, such as DNA concentration, time of reaction, α-amanitin sensitivity and re-initiation effects have been optimized some time ago, and are believed to be independent of the extract used. Therefore, I do not advocate their optimization at each step.

I present here, for the most part, the original data that was used to arrive at the conclusions referred to as “optimal conditions”. The reader should be aware that no attempts have been made to optimize the buffer conditions (salt, pH, etc.) beyond what is known to be optimal for transcription by RNA polymerase II. There is some evidence suggesting that particular promoters work better under slightly different conditions, presumably leading to stabilization of transcription factors (or their interactions) rather than RNA polymerase II itself.

1 Protein Concentration

Figure 17 shows the rate of *in vitro* transcription from three different promoters, as a function of protein concentration. It is clear that transcription from all of these promoters is optimal at an intermediate protein concentration. At lower protein inputs, transcription is inefficient, probably due simply to the low rate of formation of pre-initiation complexes. Pre-incubation of these reactions for longer periods, however, does not result in drastic improvements in the rate of transcription. On the other hand, when the protein concentration is too high, low transcription rates are believed to be due to the effect of either specific or non-specific

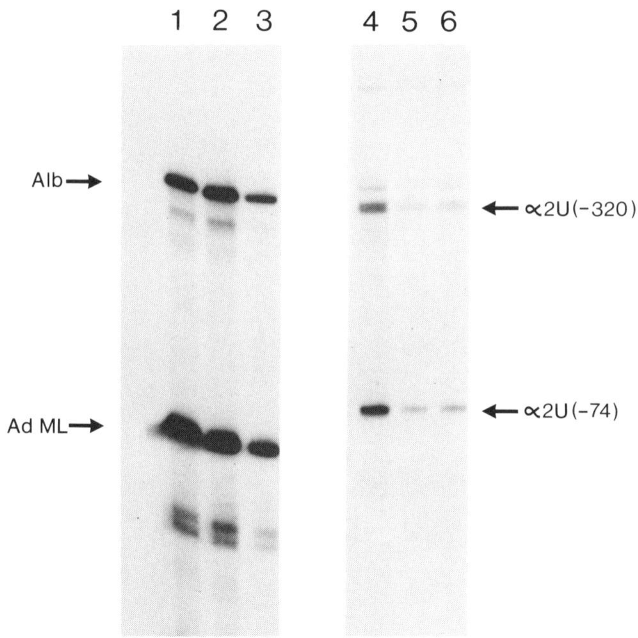

Figure 17.
Effect of Protein Concentration on the *in vitro* Transcription Reaction. Four different templates were assayed at three different concentrations of liver nuclear extract. Lanes 1 and 4: 2 mg/ml of protein, Lanes 2 and 5: 3 mg/ml of protein, Lanes 3 and 6: 4 mg/ml of protein. The templates used are indicated next to their corresponding transcripts. Notice that, at 3 mg/ml of protein, the albumin promoter works optimally, the Adeno Major Late promoter works as well as with 2 mg/ml of protein, while the α2U-globulin promoter is virtually inactive, its optimal performance being at 2 mg/ml of protein. In all cases, transcripts were assayed directly, by using the G-free cassette.

repressors present in the extracts. Binding of these molecules to either the promoters or the transcribed regions seem to interfere with initiation and/or elongation of transcription. The important lesson from this experiment is that as a rule with *in vitro* transcription, more is not necessarily better.

2 DNA Concentration

The effect of template concentration on *in vitro* transcription is, in my view, a little more subtle and interesting than the effect of protein concentration. In this case, there are two related, but separable variables to be considered: total DNA concentration, and specific template concentration. Figure 14 shows the effect of these on a reaction containing varying amounts of a template carrying the albumin promoter, either with or without further addition of carrier DNA (high molecular weight salmon sperm DNA). The results indicate that transcription from the albumin promoter is already detectable when only 6 ng of template per reaction are used, but this is only true when the total DNA concentration is kept at 800 ng/reaction, through the addition of salmon sperm DNA. If no salmon sperm DNA is added, 400 ng of template DNA are required to produce a detectable level of *in vitro* transcription. This figure also shows that the specific signal derived from the albumin promoter increases linearly with template concentration all the way up to the highest point tested, 800 ng/reaction. Additional experiments have shown that the signal plateaus around this concentration, and it stays constant until a concentration of approximately 1.3 μg/reaction. This is the basis for using 1 μg/reaction as a standard DNA concentration.

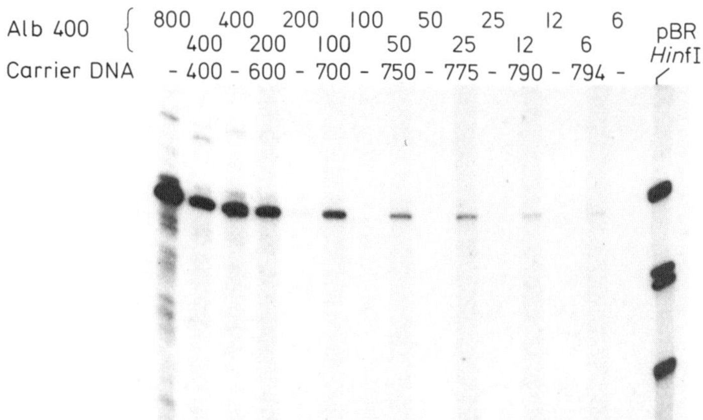

Figure 18.
Effect of DNA Concentration on the *in vitro* Transcription Reaction. Conditions were as described, except that the concentration of specific template (G-free cassette under the control of the albumin promoter) ranged from 800 ng to 6 ng per reaction. Assays were performed using decreasing amounts of specific template, with or without the addition of high molecular weight salmon sperm DNA to a total DNA concentration of 800 ng per reaction. In all cases, the protein concentration was 3.4 mg/ml. (Gorski et al., 1986).

One possible (but by no means the only) explanation for these results is that, as previously stated, the nuclear extracts contain repressor molecules (probably non-specific), which must be titrated out by the presence of relatively high concentrations of DNA, before any transcription can take place. In this regard, it is important to notice that, if the total DNA concentration is kept constant, transcription from the albumin promoter increases linearly with template concentration, in spite of the fact that very few (1–2%) of the template molecules are utilized in any given reaction. This phenomenon argues against specific repressor molecules being at the origin of the need for carrier DNA when low template concentrations are used. Experiments in which pUC19 plasmid DNA has been used instead of salmon sperm DNA indicate that eukaryotic DNA sequences (or promoters) are not necessary in order to remove the block to transcription. The fact that only 1 to 2% of the template molecules are ever utilized to produce transcripts suggests the presence of cooperativity in transcription factor binding to the promoters. Once a “nucleation” or rate limiting interaction has occurred in a particular template molecule, the rest of the transcriptional machinery will cooperatively be attracted to that same molecule, thus ensuring the formation of a successful pre-initiation complex. While this appears to be the most rational explanation, hard evidence on cooperativity of binding within the transcriptional machinery has been difficult to obtain directly and unambiguously.

3 Time

Figure 19 shows the time course of a typical *in vitro* transcription reaction. The components have been pre-incubated on ice for 10 minutes before the addition of NTPs, as described in the methods section. The accumulation of transcripts occurs linearly after a short lag, and for a period of approximately 45 minutes, after which little de novo synthesis is apparent.
In the experiment shown in Figure 19, re-initiation was allowed to proceed normally, since the conditions used are exactly those used in routine assays for *in vitro* transcription. The effect of adding heparin to the reaction at various times is shown in Figure 20. Heparin, used at a concentration of 1 mg/ml, inhibits re-initiation of transcription, while having no effect on transcript elongation. This is due to binding to the ß' subunit of RNA polymerase. The results show that transcript initiation continues to occur throughout the incubation period. This experiment suggests that formation of active pre-initiation complexes is not restricted to the 10-minute pre-incubation period at the beginning of the reaction. Nevertheless, better results are routinely obtained when the pre-incubation step is included, even if the time of incubation at 30°C is increased to 55 minutes.

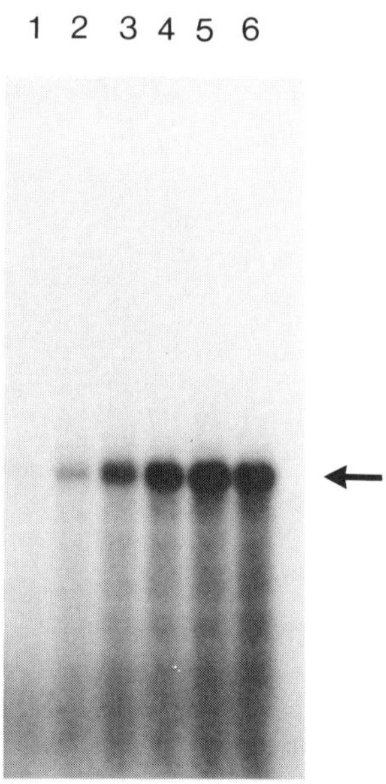

Figure 19.
Time Course of the *in vitro* Transcription Reaction. The template used contained the G-free cassette under the control of the Adeno Major Late promoter. Reactions were stopped after 10, 20, 30, 45, 60 and 75 minutes, and transcripts were analyzed by the standard method, except that a glyoxal gel was used to electrophoretically display the transcripts.

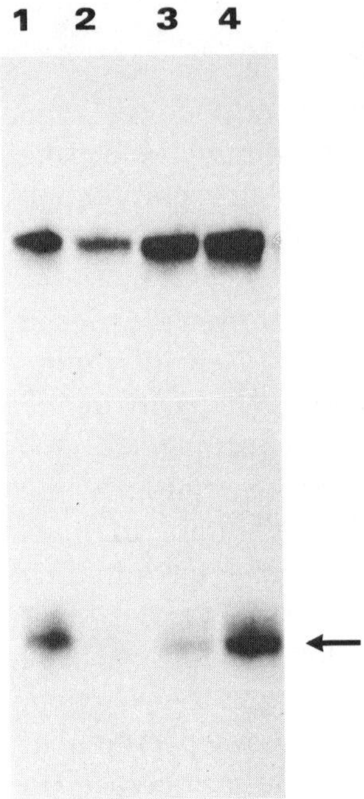

Figure 20.
Effect of Heparin on the *in vitro* Transcription Reaction. *In vitro* transcription reactions using liver extracts and a template containing the α2U-globulin promoter were analyzed by S_1. Heparin (1 mg/ml) was added 5 minutes (Lane 2), 15 minutes (Lane 3) or 45 minutes (Lane 4) after transcription was started, and the incubation was allowed to proceed for a further 30 minutes. Lane 1 is a control with no heparin added.

4 α-Amanitin

Most of the assays done with nuclear extracts prepared as described deal with transcription processes normally carried out *in vivo* by RNA polymerase II. For this reason, it was of primary importance to determine whether, *in vitro*, it was this same enzyme that was responsible for the observed transcription. In order to determine which RNA polymerase was involved, *in vitro* transcription reactions were performed that contained a low concentration of α-amanitin. This compound is widely used to distinguish between the different eukaryotic RNA polymerases. It has long been established that RNA polymerase II is inhibited by very low doses of this substance (in the order of a few μg/ml), while RNA polymerase III is usually only inhibited by much higher doses (in the order of a mg/ml) and RNA polymerase I activity is not affected by α-amanitin at any concentration tested.

Figure 21 shows that a low concentration of α-amanitin completely abolishes transcription from the albumin promoter. This experiment indicates that the enzyme responsible for *in vitro* transcription from the albumin promoter is indeed RNA polymerase II. Since the assay shown in Figure 21 was done using the G-free cassette, in a reaction that did not contain GTP, the lack of a signal in the presence of the drug only gives information about the enzyme responsible for the albumin promoter utilization. Similar experiments, but done with templates not containing the cassette, and analyzed by S_1, suggest that the extracts do contain, in addition to RNA

Figure 21.
Effect of α-amanitin and tRNA on the *in vitro* Transcription Reaction. A template containing the G-free cassette under the control of the mouse albumin promoter was transcribed *in vitro* using liver nuclear extract. Lane 1: Standard reaction, Lane 2: Same as in 1, but containing 15 μg of tRNA, Lane 3: Same as in 1, but in the presence of 2 μg/ml of α-amanitin. The transcripts were separated on a glyoxal gel, and the arrow indicates correctly initiated transcripts.

polymerase II, other forms of RNA polymerase, not sensitive to inhibition by low doses of α-amanitin. These other enzymes appear to be able of initiating transcription from pseudo-promoter sequences, probably located within vector sequences. Transcripts derived from such events contribute a negligible amount to the total transcription observed. Indeed, much of this "full-length probe" protection appears to correlate directly with the strength of the promoter tested (see, for example, Figure 24), with only a residual fraction of this signal being resistant to low concentrations of α-amanitin. This suggests that most, although not all of these transcripts are derived from RNA polymerase II starting transcription at the correct promoter, and then continuing transcription through the whole plasmid at least once.

D Tissue Specificity

The main objective we had for developing the nuclear extracts described in this book was the possibility of studying tissue specificity *in vitro*. Crucial to that purpose was the capability of these extracts to reproduce such specificity. We have tested this property in two ways: the ability of liver-specific promoters to be transcribed better in extracts from liver than in extracts from other tissues, and the inability of promoters not utilized in the liver *in vivo* to function in liver extracts *in vitro*. Figure 22 illustrates both of these points. It should be emphasized that it is not yet possible to reproduce tissue specificity *in vitro* with the degree of accuracy observed *in vivo*. Indeed, tissue specificity *in vitro* means only that transcription is more efficient in extracts derived from the expressing than from non-expressing tissues. In *in vitro* assays, transcripts are always produced, even when extracts from non-expressing tissues are used. Nevertheless, a 10 to 100-fold difference can be observed, and this is considered satisfactory for the moment. It is important to keep in mind that, *in vivo*, tissue specificity is achieved through the concerted action of several different elements, some of which are known to be inoperative *in vitro*, such as chromatin opening and very distal elements not located in the immediate vicinity of the transcription start sites.

When assessing tissue specificity of *in vitro* transcription, we have to assume that some viral promoters, such as the Adeno Major Late (AdML), work with the same efficiency in different extracts. This is more or less the case *in vivo*, but it is difficult to conclusively prove that the same is true *in vitro*. At any rate, direct comparisons between extracts derived from different tissues are only possible if control promoters can be shown to be utilized at similar rates in the extracts examined. A comparable rate of transcription from the AdML promoter might require different protein concentrations in different extracts (in fact, that is the criterion used to optimize conditions for each extract). On the other hand, if comparable signals can not be obtained, I feel it is necessary to prepare new extracts from the tissue that does not transcribe efficiently enough. I can not give a precise reason, but it is my impression that exposing films for different lengths of time in order to achieve similar signals is not quite fair, and results obtained by doing such tricks are more questionable than those in which comparable levels of expression from the AdML promoter can be obtained directly.

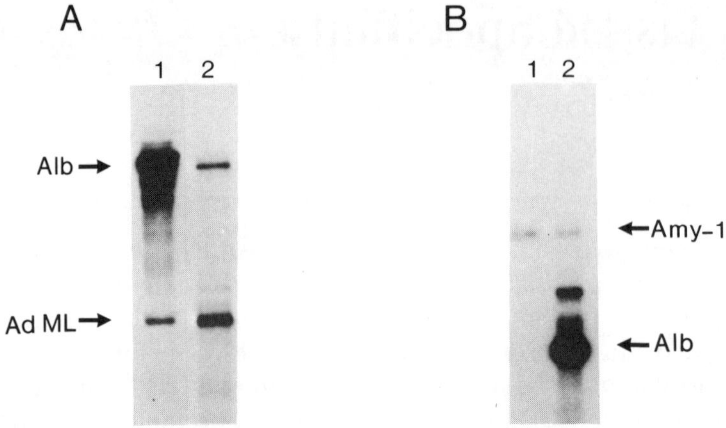

Figure 22.
Tissue Specificity of the *in vitro* Transcription Reaction. Panel A shows that a liver-specific promoter (mouse albumin) is more efficiently utilized for *in vitro* transcription in extracts from the expressing (liver, lane 1) than from non-expressing (spleen, lane 2) tissues. The non-tissue specific Adeno Major Late promoter is similarly utilized in both extracts. Panel B, on the other hand, shows that liver extracts utilize a liver-specific promoter (mouse albumin, lane 2) more efficiently than a promoter not normally expressed in this tissue (the parotid-specific Amy-1 promoter, lanes 1 and 2). (Maire et al., 1989 and Gorski et al., 1986).

Finally, it is sometimes rather difficult to establish whether a good level of transcription is being achieved if only extracts from expressing tissues are used (for example, if a promiscuous promoter is being studied). Since most promoters are transcribed at least at a low level even in non-expressing tissues, a non-regulated rate of transcription observed only in extracts from normally expressing tissues could be misleading.

E *Cis* Elements and *In Vitro* Transcription

As already mentioned, transcription *in vivo* results from the cooperation between a variety of different elements. The most important of these are chromatin accessibility, DNA modifications, enhancers, repressors and promoter elements. Many, if not all of these phenomena are the result of DNA-protein interactions, and several of these elements may play a role in the regulated expression of any given gene. While nuclear extracts are likely to contain many of the proteins involved in all of these phenomena, it is clear that not all the proteins (or other factors) are present in these soluble extracts. Therefore, only a subset of these interactions are at present amenable to study by *in vitro* methods.

Specifically, nuclear extracts prepared as described have been particularly useful in studying DNA-protein interactions relevant to promoter functions. In studying these short-range interactions, direct correlations have been made between specific mutations on a DNA element and the function of the promoter *in vivo*, *in vitro* and in footprinting experiments (see, for example, Figures 23 and 24). Interestingly, such studies have clarified DNA-protein interactions involved in both general activation and repres-

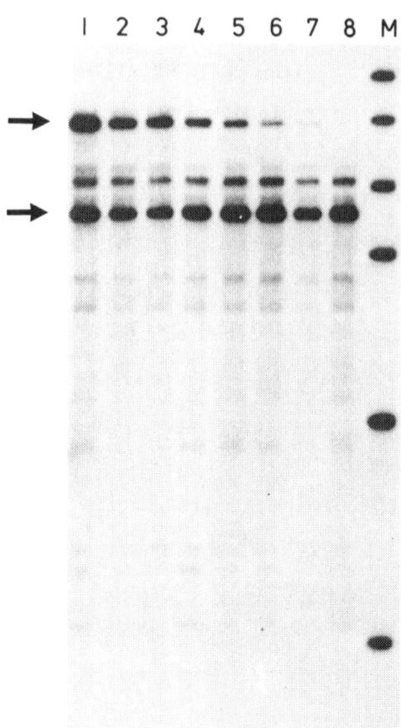

Figure 23.
Effect Of Upstream DNA Sequences On *in vitro* Transcription. Albumin Promoter. Each reaction contained an equimolar amount of an appropriate deletion mutant (upper arrow), and a standard construct (lower arrow) containing 650 bp of upstream sequences. The 5' end points of the deletions are: Lane 1: –299, Lane 2: –170, Lane 3: –147, Lane 4: –124, Lane 5: –94, Lane 6: –74, Lane 7: –64, Lane 8: – 23. (Gorski et al., 1986).

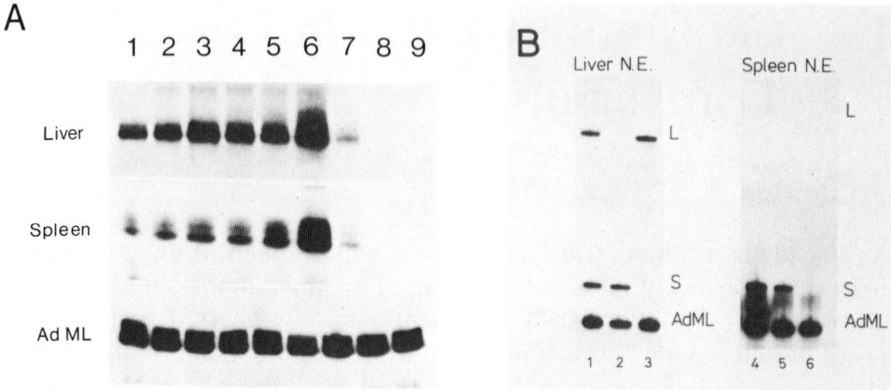

Figure 24.
Effect of Upstream DNA Sequences on *in vitro* Transcription. α-2U Globulin Promoter. A. The templates utilized were a series of 5' deletion mutants, whose end points are as follows: Lane 1: –376, Lane 2: –238, Lane 3: –182, Lane 4: –130, Lane 5: –113, Lane 6: –74, Lane 7: –43, Lane 8: –28, Lane 9: –22. The same series of plasmids was transcribed in extracts from liver and from spleen. For the series marked AdML, equal amounts of each of the α-2U globulin deletion clones were mixed with a template containing the Adeno Major Late promoter in front of the G-free cassette and incubated for *in vitro* transcription in the liver extract. α-2U globulin transcripts were analyzed by the S_1 assay. B. A similar experiment, but analyzed by the G-free cassette assay: α-2U globulin templates containing either 376 (L) or only 74 (S) bp of upstream sequences were mixed with a construct containing the Adeno Major Late promoter, and transcriptions were performed with extracts derived from either liver or spleen (Sierra et al., 1990).

sion, as well as tissue-specific activation and repression. These observations, in turn, have prompted several laboratories to isolate and/or clone these various transcription factors.

The case of enhancers and silencers, on the other hand, seems to be more complex. These elements appear to function *in vitro*, but only if located in the immediate vicinity of the promoter (as, for example, in the case of several retroviral LTR's). Distance- and position-independent activity of enhancers has not, however, been fully reproduced *in vitro*. On the other hand, *in vitro* footprinting experiments have shown that nuclear extracts do indeed contain a spectrum of proteins capable of binding to these elements *in vitro*. The fact that some of these interactions are inducible, both *in vivo* and *in vitro*, further suggest that they are indeed relevant to enhancer function. The most likely conclusion from these experiments appears to be that enhancer activity is mediated by a series of interactions between the enhancer and soluble proteins present in the nuclear extracts; on the other hand, if an enhancer is to work at a long distance from the

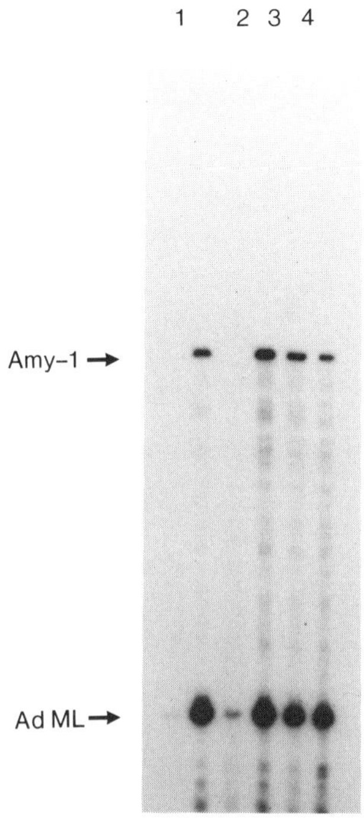

Figure 25.
Effect f Upstream DNA Sequences on *in vitro* Transcription. SV 40 Enhancer. A series of templates containing the parotid-specific promotor of Amyl-1 under the effect of the SV 40 enhancer were tested for transcription in liver extracts. Lane 1: Amy-l promoter alone, containing sequences between –64 and +6, Lane 2: Amy-1 containing the SV 40 enhancer upstream from position –64, in the "—" orientation, Lane 3: Amy-1 containing the SV 40 enhancer downstream of the G-free cassette, in the "+" orientation, Lane 4: Amy-1 containing the SV 40 enhancer downstream of the G-free cassette, in the "—" orientation.

activated promoter, this effect seems to require the participation of as yet undescribed insoluble (or extremely labile) nuclear factors. These types of factors are not present in nuclear extracts prepared as described in this book.

Finally, as it has already been discussed, soluble nuclear extracts are not able to reproduce chromatin effects. This might be due in part to the fact that templates are always small pieces of DNA, not necessarily associated into nucleosomes, and certainly unable to acquire higher order structural features. Little is known about the mechanisms of chromatin opening, so that the presence or absence of relevant factors in the nuclear extracts is difficult to assess. It is very likely that these elements will represent a situation analogous to that found with enhancers: many of the proteins involved in interactions with putative "chromatin openers" are likely to be present in the soluble extracts, but probably interactions with insoluble matrix components will be necessary to reproduce the activity of these elements *in vitro*.

IV Condensed Protocols for Benchtop Use

A Preparation of *In Vitro* Transcription Extracts

Protocol given for 12–15 g of rat liver

1) Dissect tissue and keep it in a cold beaker on ice until all animals have been dissected. Mince finely with scissors.
2) Homogenize the tissue by 4 or 5 strokes in a motor-driven Potter-Elvehjem homogenizer. For this, the tissue is resuspended in homogenization buffer at about 1 g per 10 ml.

The actual volume used depends also on the number of SW 27 tubes that will be used in the next step.

3) Load on top of 10 ml pads of the same buffer, on SW 27 tubes. The pads should be prepared in advance, with a solution containing the DTT and TRAS, but not the PMSF, which is added at the last minute. Each tube should be loaded with 28 ml of extract (38 ml volume with the pad). There are 6 buckets in the SW 27 rotor.
4) Spin at 24 K rpm for 60 minutes in a pre-cooled centrifuge (–2°C).
   Whole cells float
   Nuclei with tags stay at interface
   Clean nuclei pellet.
5) With a spatula, remove the solid disc that floats, then carefully aspirate off the liquid (with 5- or 10-ml pipet, not with Pasteur), wipe off and wash the sides of the tube extensively with water (from a long bent needle attached to a 50-ml syringe).
6) If a second spin is necessary, resuspend in homogenization buffer containing an additional 10 % of glycerol, by bubbling and scraping the side of the tube with a plastic pipet, then, aspirate up and down several times; transfer to a Potter-Elvehjem and homogenize again (only once or twice this time, and with the motor set at a lower speed).
7) Layer again onto a 10-ml pad in SW 27 tubes, as in 3.
8) Spin at 24 K rpm for 60 minutes.

Note: The Pelleted Nuclei can be stored at –70°C

9) Aspirate off the liquid and resuspend the nuclei in lysis buffer, again by bubbling and scraping with a plastic pipet, but this time followed by homogenization in a glass homogenizer fitted with an A pestle.
   The volume of lysis buffer used depends on the tissue.
10) Measure A_{260} of a 1:50 dilution in 0.5% SDS (blank against a 1:50 dilution of lysis buffer). Calculate from this the concentration of DNA.
11) Dilute with lysis buffer to have a DNA concentration of 0.5 mg/ml.
12) Add 1/10 volume of 4 M $(NH_4)_2SO_4$ and mix gently by inversion.
13) Leave in ice-water for 30 minutes, with occasional mixing. The solution should stay clear.
14) Spin for 60 minutes at 35 K rpm in the Ti60 rotor
    or at 40 K rpm in the Ti50 rotor
    At this step, it is important to remove the tubes as soon as the centrifuge stops, since otherwise the chromatin will start re-swelling.
15) Transfer the supernatant to a Ti60 or Ti50 tube (with a plastic pipet).
16) Add 0.3 g of solid $(NH_4)_2SO_4$ per ml of supernatant.
17) After all the salt has dissolved, leave on ice-water, shaking gently, for 20–60 minutes.
18) Spin for 20 minutes at 35 K rpm in the Ti60 rotor or at 40 K rpm in the Ti50 rotor.

Note: Pellet can be stored overnight in ice-water

19) Resuspend in dialysis buffer. (The volume depends on the DNA measurement done on step 10). Usually, 1 mg of DNA indicates 1 mg of protein. Extract should be resuspended at about 10 mg/ml of protein.
20) Dialyze 2 × 2 hours in the cold against 100 volumes each time of dialysis buffer. A white powdery precipitate should form.
21) Remove dialysate and spin 2 minutes in a cold Eppendorf centrifuge.
22) Aliquot and freeze immediately in liquid nitrogen.

Measure A_{230} and A_{260} of a 1:100 dilution of extract, made in H_2O, and blanked against a 1:100 dilution of dialysis buffer.
$187 \times A_{230} - 81.7 \times A_{260} = \mu g/ml$

B In Vitro Transcription

Prepare the following mix (per sample):

4 µl 5X mix
0.7 µl (α-^{32}P) UTP
1 µl RNasin (Promega)
3 µl nuclear dialysis buffer
1.3 µl H_2O.

Note: The dialysis buffer should be omitted if the reaction is to contain 60% extract.

- 9 µl extract (45% of total volume; can use 12 µl, or 60% instead)
- 1 µl template DNA (1 µg)
- Incubate on ice for 10 minutes
- Add 10 µl of the mix described just above
- Incubate at 30°C, 45 minutes
- Add a mix of:
  280 µl STOP buffer
  2 µl 10 mg/ml tRNA
  4 µl of 10 mg/ml Proteinase K
- Incubate at 37°C, 30 minutes
- Extract once with phenol-chloroform
- Precipitate with 750 µl of ethanol
- Leave in dry ice, 15 minutes
- Spin 10 minutes in the microfuge
- Resuspend pellet in 5 µl formamide dye
- Heat to 90°C for 2 minutes and load 3 µl on 4% polyacrylamide/urea gel.

C Footprinting

Preparation of the probe: The DNA fragment to be analyzed (usually between 100 and 800 bp) should be end-labelled by Klenow (or kinase) and resuspended in H_2O at a concentration of about 1–3 ng/µl. This should provide with approximately 10,000 to 100,000 cpm/µl.

On ice, mix the following:

- 13 µl of nuclear extract (plus dialysis buffer)
- 2 µl 10 × Footprint Buffer
- 0.5 µl poly (dI:dC), 2 mg/ml
- 1 µl probe (1 ng)
- 3.5 µl H_2O.

Notes:

1) All these components, except for the protein extract, can be premixed at the start of the experiment.
2) Protein concentration should not exceed 5 µg/µl (100 µg total) and usually 20–40 µg of total protein is enough.

Incubate the above mix on ice for 15 minutes.

In the meantime, prepare dilutions of DNase in the appropriate buffer and let sit on ice for 5 minutes:

- 1: 200 for Control
- 1: 100 for ~ 40 µg protein
- 1: 50 for ~ 80 µg protein.

Add 3 µl of DNase to each sample
Incubate on ice for 5 minutes
Add 80 µl of STOP solution
- + 1 µl ssDNA
- + 1 µl proteinase K

Incubate at 45°C for 60 minutes
Extract twice with phenol/chloroform and ethanol precipitate
Resuspend in 5 µl of formamide dye
Heat to 90°C for 2 minutes and load 3 µl on a 6% polyacrylamide/urea gel.

V Trouble-Shooting Guide

The following list is intended as a starting point to try to resolve potential problems that might be encountered while working with the protocols described in this book. No trouble-shooting guide can ever replace solid judgement and a thorough understanding of the biological material at hand. Therefore, it would be unrealistic to hope to present an exhaustive list of all possible problems, causes and solutions. Nevertheless, I present here the most common ones found by beginners.

A Nuclei Preparation

Problem 1: ***Insufficient Homogenization***

Probable causes and solutions:

a. The clearance between the pestle and the homogenizer might be too wide. Try interchanging several homogenizers and pestles to find the best combination.

b. Some tissues, such as muscle and kidney, contain large amounts of fibrous material and are therefore difficult to homogenize. Try mincing the tissue to a paste before homogenization (avoid however taking too long a time for this). Alternatively, a higher speed during the homogenization might be required.

c. It is important that the homogenization be performed with a motor-driven apparatus. Even the strongest athlete will fail to obtain a good homogenization if it is done by hand.

d. The volume used during the homogenization should not be such that a portion of the preparation stays above the working space of the pestle. It is better to do several homogenizations using lower volumes, or to use the modified food processor shown in Figure 4.

Problem 2: ***Low Yield of Nuclei***

Probable causes and solutions:

a. The cells might not have lysed. This could be due to a variety of factors, the most frequent of these being an incomplete or insufficient homogenization (See Problem 1). Alternatively, it is possible that the cells will fail to lyse if they contain unusually small cytoplasmic volumes. I am not sure what could be done in this case, but I would guess that stronger (or longer) homogenization would be less deleterious than modifying the ionic strength of the homogenization buffer.

b. The nuclei themselves might have lysed. This would be evident because of the presence of a large interface between the cushion and the upper phase. Again, the most likely problems are shearing, caused by too fast a homogenization speed, and/or a bad clearance between pestle and homogenizer (too small in this case). Nuclei will also lyse if the homogenization time is too long. All these variables have to be optimized simply by practice, but unless extreme conditions are used, only partial lysis will result, so that usable extracts will still be obtained in the meantime...it just gets better with time, practice and solid judgement.

c. The sucrose cushion might be too dense for the nuclei in question. Nuclei isolated from different tissues or animals might differ in their density and/or size. Do not use more than 2 M sucrose in the cushions. If this is perceived as a potential problem, try less sucrose in the cushions, or longer spinning times. The nuclei will also fail to enter the sucrose cushion if large cytoplasmic tags are still attached to them. Again, in this case they will remain at the interface. Try longer homogenization or tighter pestle combinations (See Problem 1).

d. Of course, there is the obvious possibility that you are working with a very small tissue, that does not contain a large number of nuclei. Try scaling down the whole protocol, or, if feasible, pool a larger number of animals to do the experiment (again, avoid processing so many animals that the dissection step becomes too long).

Problem 3: *Nuclei Appear Dirty*

Probable causes and solutions:

a. If the pellet appears pinkish or brownish, the most likely reason is contamination with passenger red blood cells or intracellular debris. Try to avoid allowing excessive blood to contaminate the tissue during dissection; a very short wash into a beaker containing cold PBS before mincing is advisable. Otherwise, a second step of homogenization and centrifugation becomes necessary, since these preparations have been found to perform very poorly in transcription assays.

b. Sometimes, large cytoplasmic tags remain associated with the nuclei. These will not necessarily appear dirty to the naked eye, and careful microscopic examination is always advisable. Occasionally, cytoplasmic tags will prevent the nuclei either from entering the sucrose cushion, or from forming a tight pellet at the bottom of the tube. Again, the only sensible solution is further homogenization and a second high speed spin. It becomes repetitive, but nuclei with cytoplasmic tags also perform poorly in transcription assays.

c. Otherwise clean nuclei could become dirty if the sides of the tube are improperly washed after centrifugation and removal of the supernatant. Always keep the tube upside down after sucking off the liquid, wipe the sides well with a Kleenex, and wash extensively with plenty of distilled water. Wipe the sides dry afterwards, so as to keep residual water from lysing the nuclei in the pellet.

Problem 4: ***Material Stays Trapped at Interface***

Probable causes and solutions:

a. Too much tissue was homogenized in a small volume of buffer. If too much tissue is homogenized at once, the resulting shearing forces (and maybe associated osmolarity changes) tend to induce a massive breakage of nuclei. The resulting chromatin will form a mesh at the interface with the cushion, and this mesh will trap most of the remaining intact nuclei, giving rise to a very low yield of material. Try using less starting material. Also, when tissues with too much DNA are used (such as calf thymus, for example), the excess of DNA can soak the polyamines, leading to EDTA-induced nuclear lysis. If this is perceived as a potential problem, increase the concentration of spermine and spermidine in the homogenization buffer

b. The nuclei might lyse for reasons other than overloading, such as a small clearance of the pestle and/or doing the homogenization step at too high a speed. See Problems 1 and 2.

c. If the sucrose cushion is too dense and/or the nuclei are too light, intact nuclei will fail to enter the cushion, and will remain at the interface. Do not use more than 2 M sucrose in the cushions. See Problem 2.

B Extract Preparation

Problem 5: ***Low Absorbance at 260 nm***

Probable causes and solutions:

a. If the A_{260} is too low in spite of the presence of a good size nuclear pellet, it is possible that the nuclei were not fully resuspended. Resuspend the nuclear pellet by vigorously pipetting up and down with a plastic 10-ml pipet. Avoid the use of glass pipets, since nuclei stick more readily to these. Also, try wetting the side of the pipet with buffer before touching the nuclear pellet, so as to minimize adherence of nuclei to the pipet.
b. The aliquot might not be representative. Remember that you are dealing with a suspension, rather than a solution, therefore, it is crucial that the suspension be homogeneous. Use a glass Douncer to resuspend clumps of nuclei, and take your aliquot right away, before the nuclei start settling.
c. Even if the nuclear aliquot is homogeneous, it is important to have a good, homogeneous lysis of the nuclei before reading the A_{260}. Add the aliquot of nuclei to the SDS while vigorously vortexing, and keep vortexing for a minute or two afterwards. It might be a good idea to pipet the whole solution up and down a few times while vortexing (with a P1000 attached to a blue tip). After transferring the solution to the quartz cuvette, be certain to allow trapped air bubbles to regain the surface.

Problem 6: ***Chromatin Swelling***

Probable causes and solutions:

a. If the lysis step is done too roughly, DNA breakage will occur, and this will result in a looser chromatin pellet. Similarly, DNA might get partially cleaved when working with tissues exceedingly rich in DNases. Be sure enough EDTA is present during the homogenization step (large amounts of tissue might carry enough Mg^{+2}), and be gentle while mixing the solution after adding the $(NH_4)_2SO_4$.

b. As soon as the centrifugal force is no longer applied, chromatin will start re-swelling from the bottom of the tube. Use full brake while the centrifuge is coming down and be ready to take the tubes out and process them as fast as possible once the centrifuge stops.

Problem 7: ***Low Final Protein Concentration***

Probable causes and solutions:

a. The A_{260} reading before lysing the nuclei might have been inaccurate due to incorrect resuspension (clumps), or the presence of air bubbles (from the SDS and the DNA). See Problem 5.

b. Some tissues, such as spleen and liver from aged animals, contain an "abnormal" DNA/protein ratio, so that a 1:1 correlation can not be used. When working with a tissue you do not have experience with, resuspend the final protein pellet in 1/2 or 1/3 of the volume suggested for liver extracts. More often than not, this gives a good protein concentration, and it is rare that the final concentration will be so high as to produce a problem (such as protein aggregation, which might occur at concentrations higher than 10 mg/ml).

c. If the homogenization buffer was not cold enough, it is possible that some leakage of proteins might have occurred during the first steps of the protocol.

d. The proteins might not have been fully resuspended. After the $(NH_4)_2SO_4$ precipitation, the protein pellet appears white, and it becomes transparent almost immediately after adding the dialysis buffer. This does not mean that the proteins are already in solution. It is wise to mark the position of the pellet with a marker pen, and keep

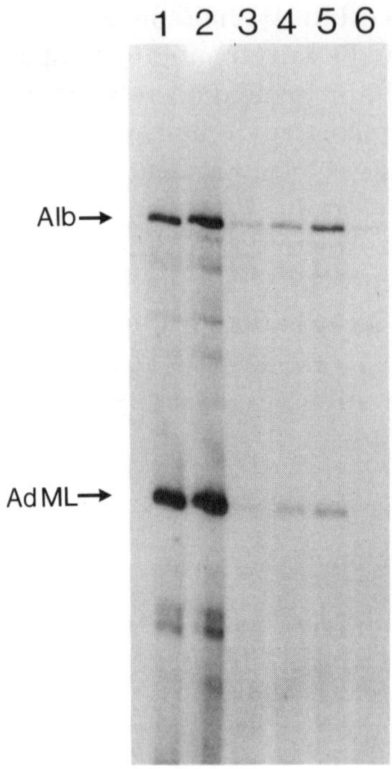

Figure 26.
Effect of Low Protein Concentration on *in vitro* Transcription. Two different liver extracts were tested for *in vitro* transcription from the mouse albumin and the Adeno Major Late promoters. Lanes 1 and 4 contain 6 μl of nuclear proteins, Lanes 2 and 5 contain 9 μl, and Lanes 3 and 6 contain 12 μl of nuclear proteins. The extract used in lanes 1 through 3 contained 7.8 mg/ml of protein, while the extract used in Lanes 4 through 6 contained 4.0 mg/ml of protein. Notice that Lanes 1 and 6 contain roughly the same final protein concentration, yet they differ markedly with respect to their transcription ability.

resuspending (gently and in the cold) for about 30 to 60 more minutes, so as to be sure all proteins are in solution. Be careful to avoid bubbling, since this tends to denature many proteins.

e. Apparently, extracts with low protein concentration can be salvaged by passing through a small heparine-agarose column. I have never tried it, so I will not give a protocol. It should also be pointed out that other assays, such as footprinting and gel retardations, require less protein than *in vitro* transcription, and extracts with low protein concentration are still useful for these other assays.

Problem 8: ***Abnormal A_{230}/A_{260} Ratio***

Probable causes and solutions:

a. Normally, good nuclear extracts give an A_{230} that is about 4 to 6-fold higher than the A_{260}. If the A_{260} is too high, that reflects the presence of contaminating nucleic acids (both DNA and RNA). Unfortunately, it appears that these are able to complex with histones during the dialysis, which means that these proteins get carried over as well (most of the histones should precipitate during the dialysis). To avoid carrying over nucleic acids, see Problem 6. If the A_{260} is too high relative to the A_{230}, the extracts do not usually perform properly in either transcription or footprinting experiments.

b. Again, this problem could be due to excessive leaking of proteins during the homogenization step, either because of the use of a homogenization buffer that is not cold enough, or taking too much time during the homogenization step.

c. DNA breaks can be induced if the extracts are handled too roughly after the first addition of $(NH_4)_2SO_4$. Small pieces of DNA will not be removed together with the rest of the chromatin during the subsequent centrifugation step, leading to contamination of the final extract.

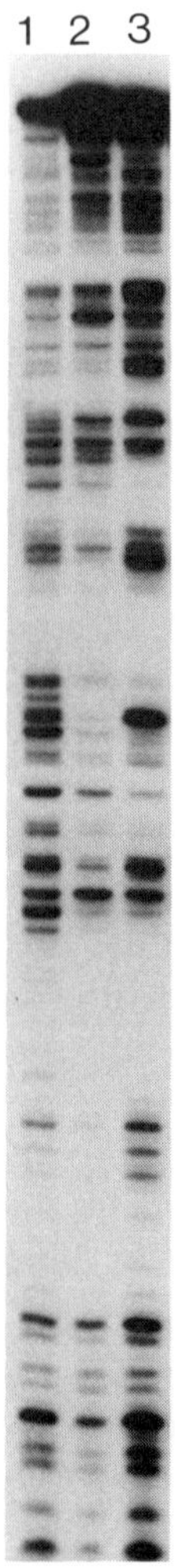

Figure 27.
Effect of a Low A_{230}/A_{260} Ratio on Footprinting Assays. Footprints on the promoter region of the α-2U globulin gene. Lane 1: Control without proteins, Lane 2: 7.4 mg/ml, extract with a ratio of 2.0, Lane 3: 4 mg/ml, extract with a ratio of 4.4.

C *In Vitro* Transcription

Problem 9: ***All Signals Are Low***

Probable causes and solutions:

a. Protein concentration is too low. Generally, good transcription is obtained with final protein concentrations in the range between 3 and 4 mg/ml (in the transcription reaction). Since the maximum volume of extract you can add is 12 μl, extracts whose initial protein concentration is below 5 mg/ml will often fail to provide enough protein factors to support vigorous transcription. See Problem 7 for tips on how to avoid low protein concentrations.

b. Even if the total protein concentration is in the right range, it is still possible that the concentration of "good" proteins is too low, due to contamination with either histones (See Problems 6 and 8) or with cytoplasm or other organelles (See Problem 3).

c. The extract and/or the template DNA might contain RNases. If that is the case, small molecular weight bands should be present. Try increasing the concentration of ribonuclease inhibitors, increasing the DTT concentration, so as to ensure full activity of RNasin, or shortening the incubation time.

d. The extract might contain proteases. In this case, a time course would show a shortened time of linearity. Increase the number of different protease inhibitors added to the homogenate.

e. Low incorporation of radioactivity might be due to a relatively large pool of UTP coming from the extract. Be sure you dialyze the extract long enough and against a sufficiently large excess of buffer. Maybe try changing the radioactive nucleotide (not all pools are identical, and specific degradation of one particular nucleotide is also theoretically possible).

f. The template DNA might be contaminated. The most frequently found contaminant is residual ethidium bromide from CsCl purified DNA. Either make sure no traces of it are left, or switch to another method of preparing the DNA, such as the one described in chapter III.B.3.1.
g. Residual RNA in the template preparation can also inhibit transcription from certain promoters (see for example Figure 21).
h. When using radioactive UTP in the reaction, it is important that it not be too old (no more than a half-life), since radiolytic products seem to inhibit the *in vitro* transcription reaction, so that the decrease in signal with respect to time is faster than predicted from decay alone.

Problem 10: ***Control is Transcribed, Gene of Interest is Not***

Probable causes and solutions:

a. A G residue is left between the cap site and the G-free cassette. If this possibility exists, check your construct by direct sequencing of the relevant junction. Also, make sure that S_1 or other assays used to determine the cap site *in vivo* were done with the accuracy required, so that transcription actually does start exactly at the nucleotide you are counting as cap site.
b. The gene might have more than one start site, and these might respond differently to the *in vitro* transcription machinery. Leaving only the upstream-most site is not necessarily enough, and it is possible that the G-free cassette is not the right assay in this case.
c. The promoter of interest might have bacterial methylation sites. Methylation at these positions might then interfere with the binding of transcriptional factors. Grow your plasmid in *dam/dcm* bacterial strains. Of course, the opposite argument is also theoretically valid: your gene might require *in vivo* methylation at CpG sites for its activity. Loss of these modifications during cloning will lead to a reduced rate of transcription *in vitro* (as well as upon transfec-

tion experiments). I do not see any easy solution to this last problem.

d. Of course, it could be simply that you are out of luck, and your promoter performs too weakly under the described conditions. You can try changing things like salt concentration, or type of ion, etc...it works sometimes. Otherwise, this is your chance to improve on our method of preparing extracts!

Problem 11: ***Background Bands Higher than Expected***

Probable causes and solutions:

a. The most common cause of this type of artifact is too low a concentration of O-methyl G during the transcription reaction. Add more O-methyl G to the transcription cocktail.

b. In most cases, the G-free cassette has been cloned into a Sma I site, which provides a GGG triplet at the end of the transcription unit. This is very effective at stopping transcription. When alternative constructs are used, it occasionally happens that a single G does not stop polymerization effectively, and a series of "stop" sites are observed. Again, try increasing the concentration of O-methyl G.

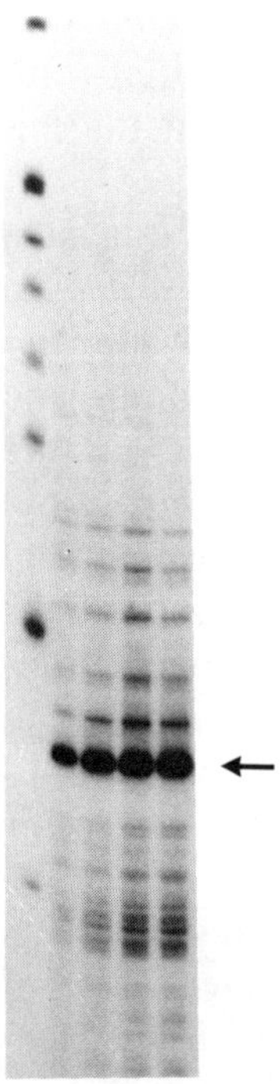

Figure 28.
Requirement for a Chain Terminator on *in vitro* Reactions Using the G-free Cassette. Several *in vitro* transcription reactions using liver nuclear extract were performed with a template containing the Adeno Major Late promoter in front of the G-free cassette. The transcription cocktail did not contain O-methyl Guanosine.

Problem 12: ***Background Bands Lower than Expected***

Probable causes and solutions:

a. Transcripts might be degraded due to the presence of RNases in the extract and/or template. Increase the amount of ribonuclease inhibitors or increase the DTT concentration, to ensure full activity of RNasin.
b. Premature termination might occur if the concentration of one or more of the nucleotide triphosphates becomes limiting. Try chasing with cold UTP. It works, but only partially.

Problem 13: ***Sample Does not Enter the Gel***

Probable causes and solutions:

a. This is usually due to the presence of proteins in the sample. Two main sources of problem can be identified: the proteinase K digestion might not be complete, or the phenol/chloroform extraction was not successful. Prepare a new stock of proteinase K and keep it in small aliquots (0.2–0.5 ml) at –20°C. If an interphase is visible after phenol/chloroform extraction, repeat the extractions until no interphase is visible. No interphase is ever visible if the proteinase K has worked properly.

Problem 14: ***Correctly Initiated Transcripts in the Absence of Template***

Probable causes and solutions:

a. This problem is possible only when transcripts are analyzed by methods other than the G-free cassette (See section III.B.3.4). Problems inherent to the S_1, primer extension or ribonuclease mapping methods will not be discussed here.
b. In the absence of template, the probe might be hybridizing with nuclear RNA derived from the promoter in question, but coming from the extract directly. Be sure that your probe contains a sufficient length of vector sequences located downstream of the start site. If using S_1 mapping, use a site located within the vector to introduce the label, if using RNase mapping or primer ex-

tension, allow for a long enough portion of vector sequences in the probe so as to be able to distinguish endogenous transcripts from *in vitro* synthesized ones.

Problem 15: ***Template-Dependent Full-Length Probe Protection***

Probable causes and solutions:

a. If the intensity of the full-length protected band parallels the strength of the band corresponding to correctly initiated transcripts, then this is most likely due to round-and-round transcription throughout the whole vector, but initiated at the strong start site provided by your promoter. There is nothing to worry about.

b. If the intensity of the full-length protected band does not change in parallel with the strength of the promoter under study, it is most likely due to non-specific start sites within vector sequences (mostly by pol I or pol III, as determined by α-amanitin sensitivity studies). Again, there is not much to be done, but I also feel that there is nothing to worry about. Alternatively, this could be due to incomplete removal of the template after *in vitro* transcription. Check the activity of the DNase.

D Footprinting

Problem 16: ***Bands are Smeared or Fuzzy***

Probable causes and solutions:

a. Too much protein during the reaction. When starting to do footprints, many people have a tendency to think that more is better. It is advisable to start by doing a protein dosage curve. In most cases, about 60 ug of protein in a 20 µl reaction (3 mg/ml) is already saturating, and, in my hands, 100 µg of protein is always too much. If you suspect that the protein giving your favorite footprint exists at a particularly low concentration in the cell, try fractionating the extract through heparin-agarose or another similar resin, so as to concentrate the protein fractions.

b. The probe might be contaminated. It is important that the probe be as free as possible of contaminating material derived either from the vector or from other bands. It is advisable to subclone the desired fragment within a polylinker, so that it can be easily cleaved free of vector. Run the preparative gel as long as possible, to obtain good separation between vector and insert, and cut the band with a clean scalpel, as tightly as possible. Clean the probe thoroughly before use.

c. Residual proteins remain after proteinase K and phenol/chloroform extraction. The most common causes are described under Problem 13. Due to the higher resolution required for footprinting experiments, it is important to do two phenol/chloroform extractions, precisely to avoid smeary results.

d. The probe is too long and/or the gel is too loose. In initial screenings for the region of interest, it is common to choose rather large fragments of DNA, which have to be run in gels containing a low percentage of acrylamide. These gels, being less resolving, necessarily give less sharp pictures. Once the region of interest has been more closely located, the use of a shorter probe (and a tighter gel) is advisable.

e. Notice that, when using total, unfractionated extracts, bands of higher molecular weight than the probe are often observed. These bands represent end-to-end ligation of the probe fragments, which happens even after short incubation times on ice. These artifacts do not seem to interfere with the quality of the footprint, and do not represent impurities in the probe.

Problem 17: ***No Footprints Observed***

Probable causes and solutions:

a. Not enough protein was present. Increase the amount of nuclear extract added to the assay. It is best to do a protein titration, expanding from 0 to a 100 μg of protein per 20 μl reaction. If this is not enough, try fractionating the extract as described under Problem 16.

b. The extract might be contaminated with nuclear DNA. See Problems 6 and 8. DNA present in the extract might non-specifically bind many of the DNA-binding proteins you are trying to assay. Similarly, the concentration of competitor poly (dI:dC) might have to be titrated. In some rare cases, poly (dI:dC) is not the best competitor, and should be replaced either by *E. coli* DNA or by another synthetic oligonucleotide.

c. Maybe the chosen DNA region does not really have specific protein binding sites. It is assumed that any region of DNA that, when deleted or mutated induces a biological activity, should contain protein binding sites. It is however possible that the effect is mediated by protein-independent processes such as DNA bending; these processes in turn could be mediated by proteins binding at distant sites, thus becoming tissue- or otherwise specific. If the biological activity has not been reproduced by *in vitro* transcription, it is also possible that the interesting DNA region does indeed contain protein binding sites, but that these proteins are lost during the preparation of the soluble extracts. Again, this is a nice opportunity to improve upon our method for preparing the extracts.

d. Maybe a protein does bind to the region, but the DNA-protein interaction is unstable. This possibility will be further discussed under Problem 21.

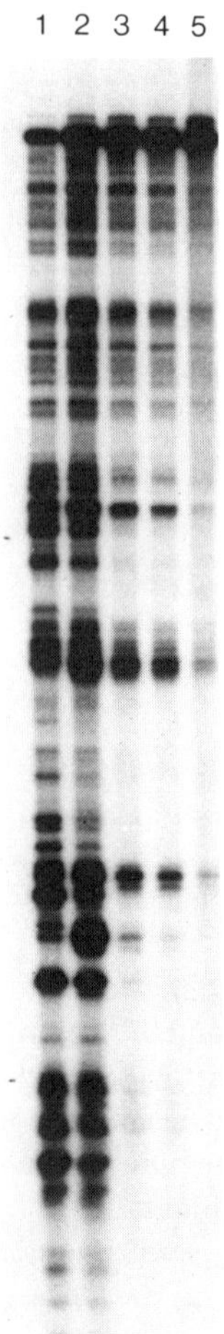

Figure 29.
Effect of Varying the Protein Concentration in Footprinting Analysis. Footprints of the promoter region of the α-2U globulin gene were performed with increasing concentrations of liver nuclear extract. At low protein imput (Lane 2), the pattern resembles the control with no proteins added (Lane 1), while at very high protein concentrations (Lane 5) a complete blank out and smearing is apparent.

Problem 18: ***Complete Blank-Out***

Probable causes and solutions:

a. The extract contains histones or other non-specific DNA-binding proteins. See Problems 6 and 8. Non-specific DNA binding proteins will cover the whole length of the probe, rendering it completely protected from DNase I cleavage. To distinguish this from the possibility of not having enough DNase I, do a control assay in which nuclear proteins are replaced by a similar amount of BSA. In this case, no footprints within an otherwise normal degradation pattern should be observed.

b. Too much nuclear extract. Since regular nuclear extracts do contain a certain amount of non-specific DNA binding proteins, adding too much

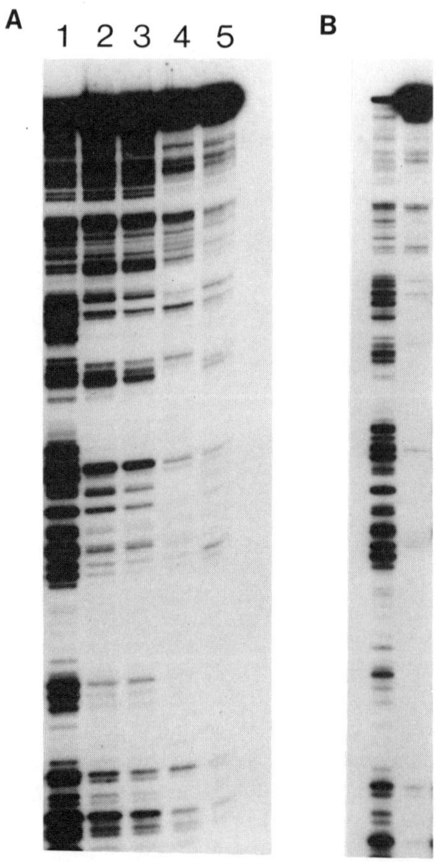

Figure 30.
Effect of Contamination with Histones on Footprint Assays. A. Footprints of the promoter region of the α-2U globulin gene were performed with two different preparations of liver nuclear extract. Lane 1 is a control with no proteins added, Lanes 2 and 4 contain 4 mg/ml of protein, while Lanes 3 and 5 contain 8 mg/ml of protein. The extract used in Lanes 4 and 5 is thought to be contaminated with histones, even though the exact nature of the contaminants has not been proven. B. Similar experiment, with yet another nuclear extract preparation.

nuclear extract to the reaction will have the same effect described under section a.

c. Titrate the amount and time of DNase I digestion. If extreme conditions are used, complete blank outs can be obtained: too much DNase I (or too long a digestion time) will degrade all the probe into small oligonucleotides that will not be retained in the gel, while too little enzyme (or too short a digestion time) will fail to cut the probe at all. To guess in which direction you should go, look at the relative amount of full-length probe.

Problem 19: ***Top of Gel is Blank***

Probable causes and solutions:

a. Too much DNase I or too long a digestion time. The amount of DNase I and time of digestion should be such that, on the average, each DNA molecule is cleaved only once. A substantial amount of undigested full-length probe should remain at the end of the experiment. The conditions required for good footprinting are in this sense analogous to those required for Maxam & Gilbert sequencing reactions. Notice that, as you purify a protein fraction, less and less mass of total protein is required to produce a footprint. The amount of DNase I used should be reduced accordingly (See the protocol for footprinting for details).

Problem 20: ***Bottom of Gel is Blank***

Probable causes and solutions:

a. Too little DNase I or too short a digestion time. Titrate the amount of DNase I, rather than the time.

Problem 21: ***Irreproducible Footprints***

Probable causes and solutions:

a. Temperature of the DNase I digestion. Many people perform the binding and/or DNase I digestion steps at room temperature or at 37°C. This is fine for most purposes, however, I feel that

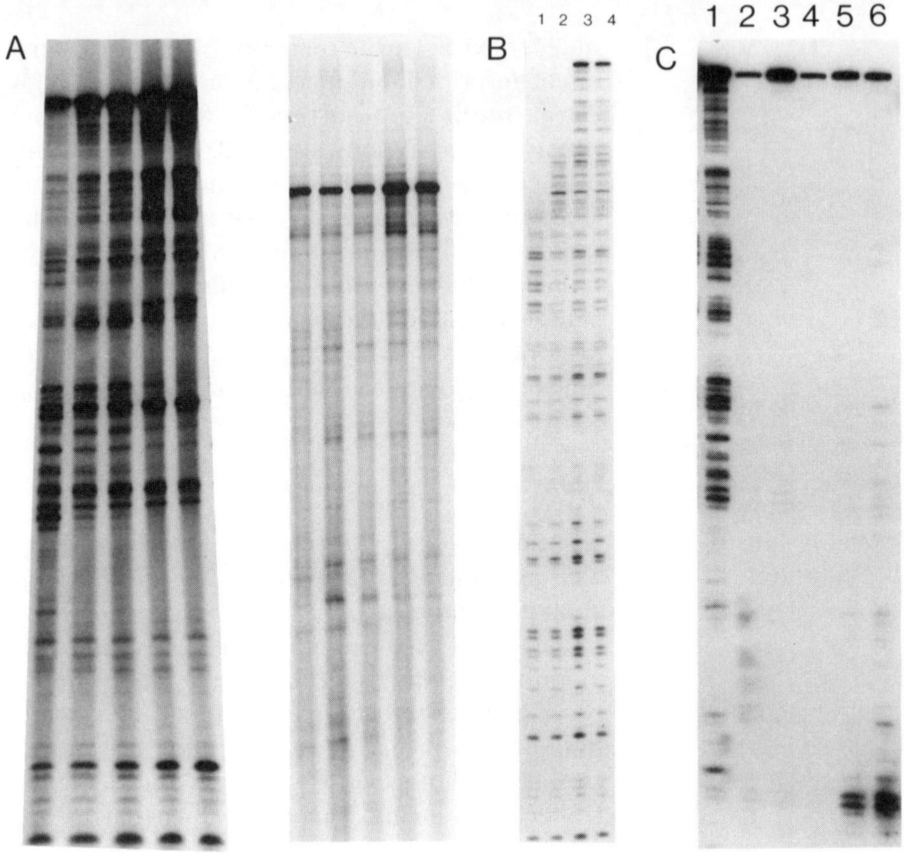

Figure 31.
Various Other Problems Encountered with Footprinting Experiments. A. In both cases, the probe was dirty. On the left, the probe was not sufficiently separated from other labeled fragments, while on the right, residual phenol was present. In panel B, too much DNase I was used (Lanes 1 and 2), while in panel C, too little DNase I was used (Lanes 2 through 6).

the reaction is best controlled if done at a lower temperature. I do all footprinting manipulations on ice.

b. The protein of interest, or a modification present on it might be labile. Again, problems are less likely to occur at lower temperatures. Also, the binding time should not be unnecessarily long. Most binding reactions are done in a few minutes

(3 or 5 minutes, which is the shortest I have cared to measure), and most interactions are relatively stable on ice; however, the binding should not be allowed to proceed for very long, just in case one of the protein components involved is not particularly stable.

c. Two proteins might compete for the same, or overlapping sites. This is a more exciting (therefore less likely) possibility. The two competing proteins might be present at different relative ratios in separately prepared nuclear extracts, or in extracts coming from different sources. Before jumping to conclusions, it is important to confirm the results. This is most elegantly done by chromatographic fractionation in the case where both proteins coexist in the same extract. If the two proteins are derived from different sources, mixing experiments could lead to displacement of one protein by another, therefore inducing a change in the observed footprints. The presence of two different proteins binding to the same site might also be inferred from gel retardation experiments.

VI Appendix: Suppliers of Special Items and Construction of Commercially Unavailable Equipment

A Chemicals

Trasylol is a trade mark for Aprotinin, sold at 1.4 mg protein/ml. It is commercialized by Bayer AG, Grubenstrasse 6, CH–8045 Zürich, Switzerland.
Sucrose should be free of ribonucleases. I use either Schwarz Mann (article number 821721) or Merck (article number 7654).
RNasin is a ribonuclease inhibitor commercialized by Promega Biotec and its European subsidiaries (in the USA: Promega Biotec, 2800 South Fish Hatchery Road, Madison, WI 53711, USA).
O-methyl G (3'-O-methyl guanosine-5'-triphosphate) is available from Pharmacia (article number P–L 27–4675–01).
Poly (dI:dC) is available from Sigma (article number P9514).

B Commercial Instruments

Regular conical glass homogenizers fitted with Teflon pestles with a clearance of 0.2 mm are from Kontes (Headquarters: Spruce Street, P.O. Box 729, Vineland, New Jersey 08360, USA).
Several motors for driving the homogenizer are available. I have used a Potter S model from B. Braun (Diessel Biotech, Postfach 128, D–3508 Melsungen, FRG) as well as a model RW 20, from Janke & Kunkel, IKA Werk, D–7613 Staufen i. Br., FRG. The settings for homogenization are: 1200 in the Braun apparatus, and 6 in the IKA Werk machine.

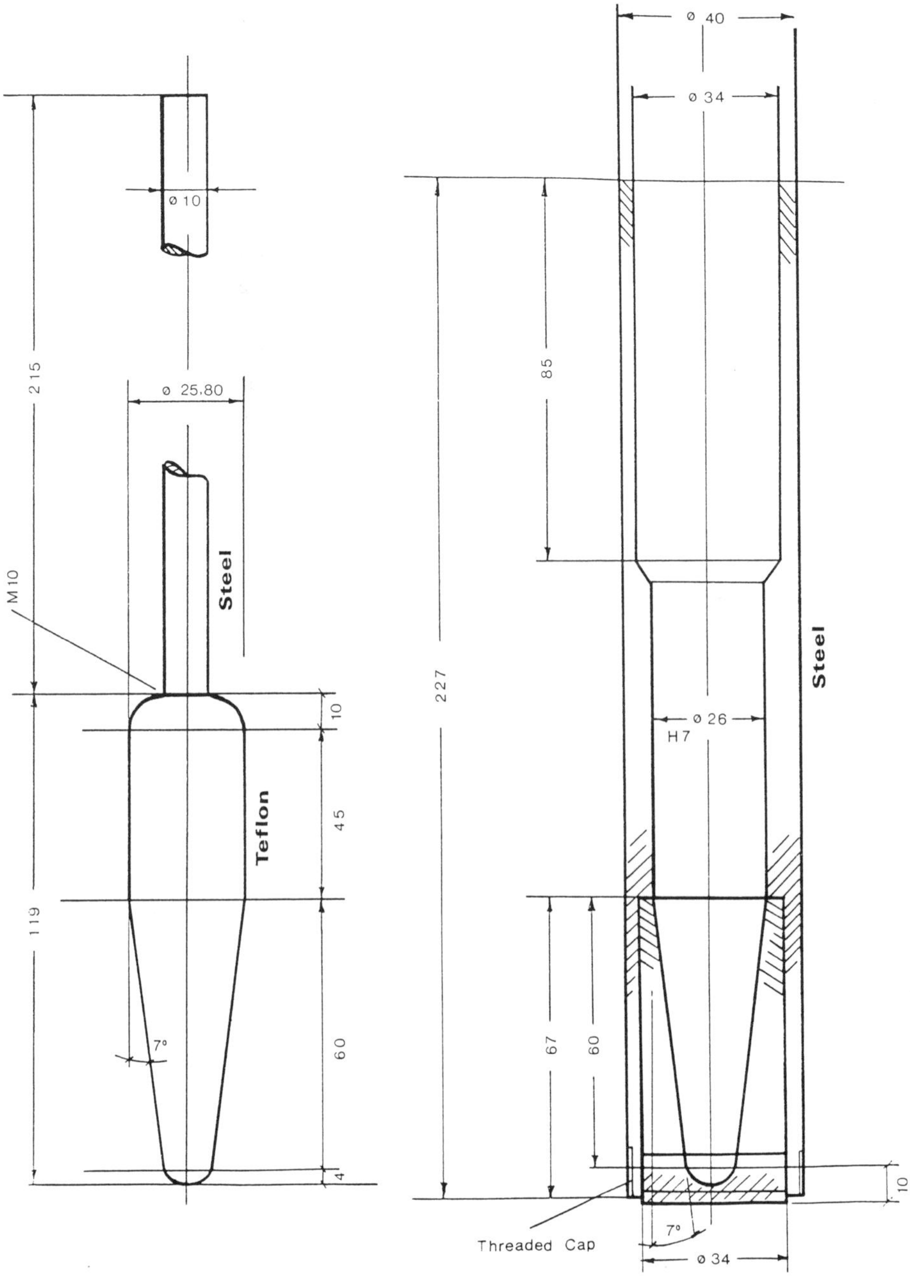

Figure 32.
Blueprint for the Construction of the "Bazooka".

C Home-Made Instruments

Bazooka homogenizer. This apparatus fulfills exactly the same function as a regular glass homogenizer. Its only advantage is that, being made out of stainless steel, it is virtually unbreakable. The high negative pressures produced inside the homogenizer when pulling the pestle out result in occasional implosions when using conventional homogenizers. Figure 4 presents a photo of the bazooka, while Figure 32 presents a blueprint for the construction of the apparatus. The pestle is from Kontes, unmodified. If the clearance is too narrow, the pestle can be shaved a few fractions of a millimeter. The tube itself, made out of four separate parts that fit tightly together, is made of stainless steel.
Modified food processor. This apparatus, also shown in Figure 4, and in blueprint form in Figure 33, was designed in U. Schibler's laboratory at the University of Geneva, and is only used when large amounts of extract are to be prepared (for protein purification purposes). The size of the container can be varied, but the important part is the lid. This is designed so as to be able to completely prevent the trapping of air within the working space, since air would lead to foaming, which denatures proteins. The motor and blades are provided by a regular household food processor, while the container and lid are made of lucite and stainless steel.

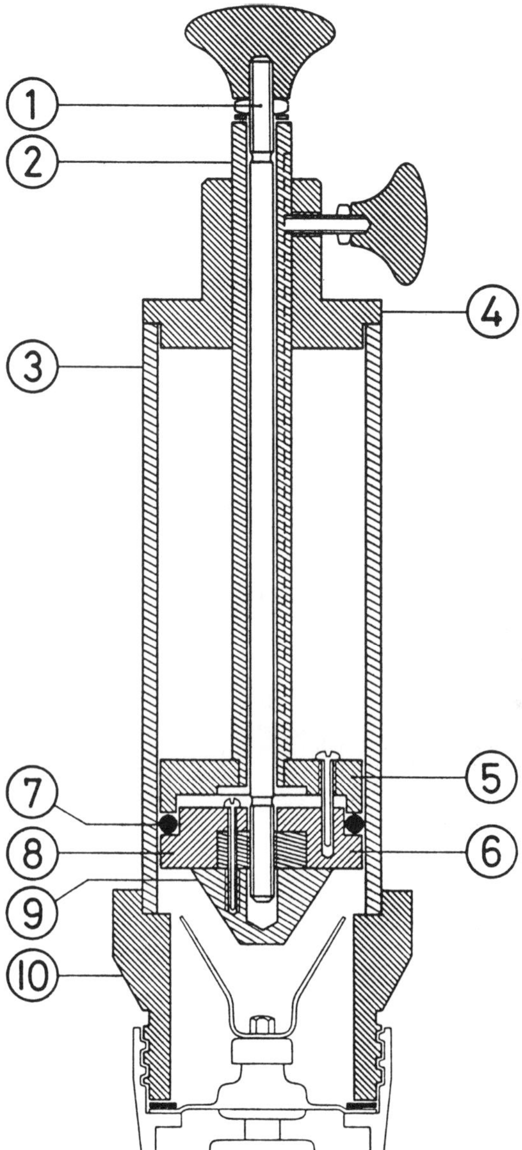

Figure 33. Part I
Blueprint for the Construction of the Modified Food Processor

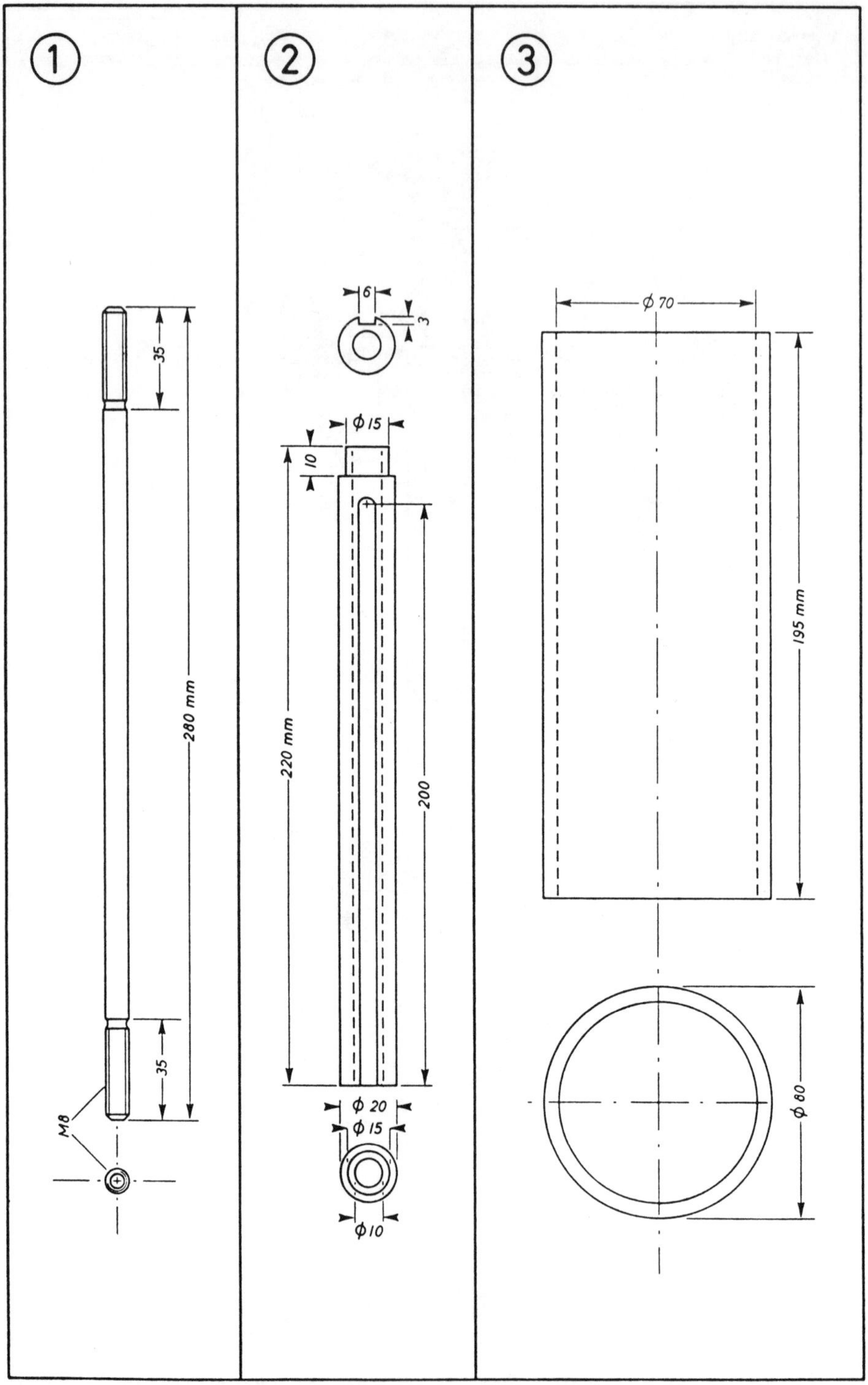

Figure 33. Part II.

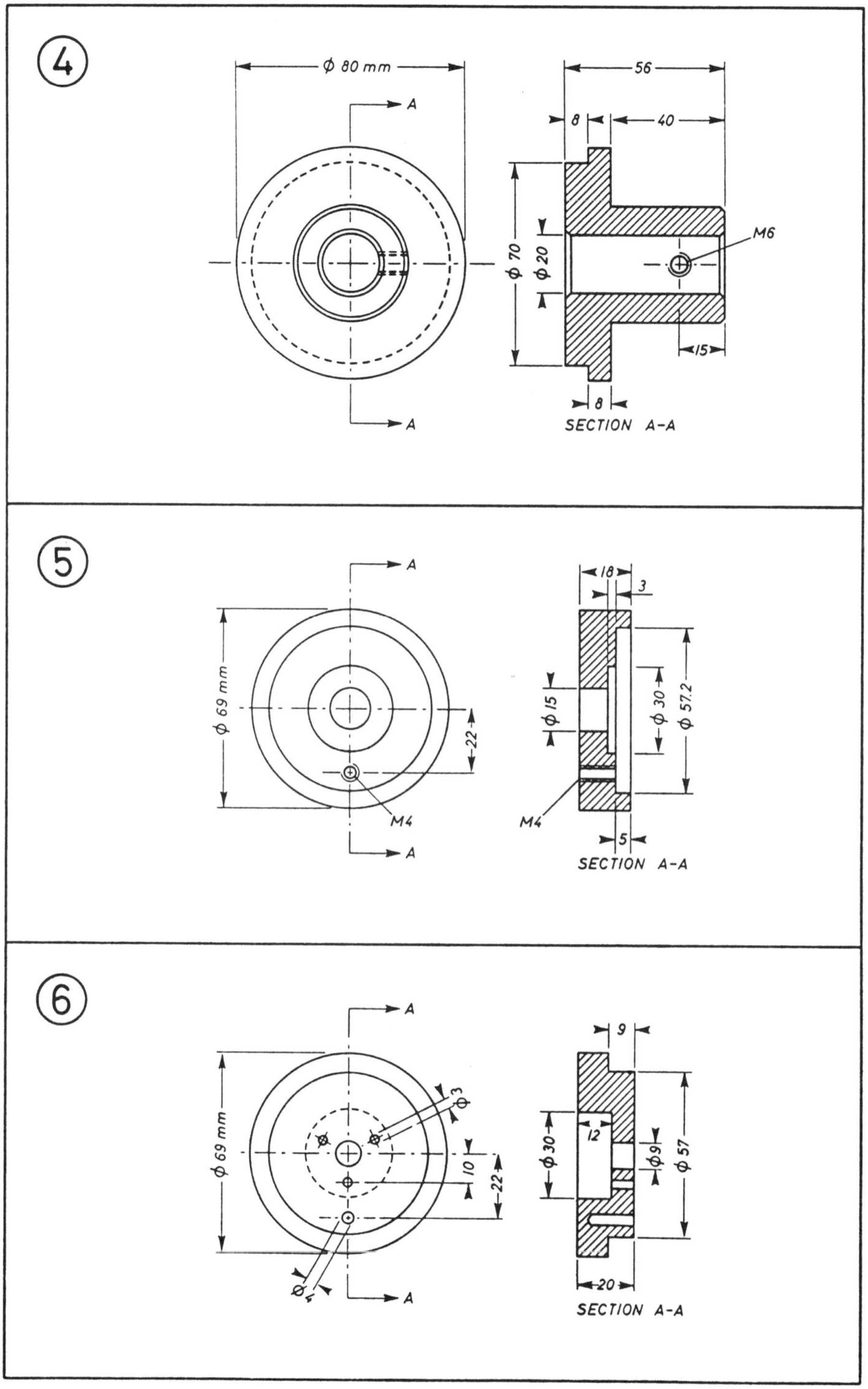

Figure 33. Part III.

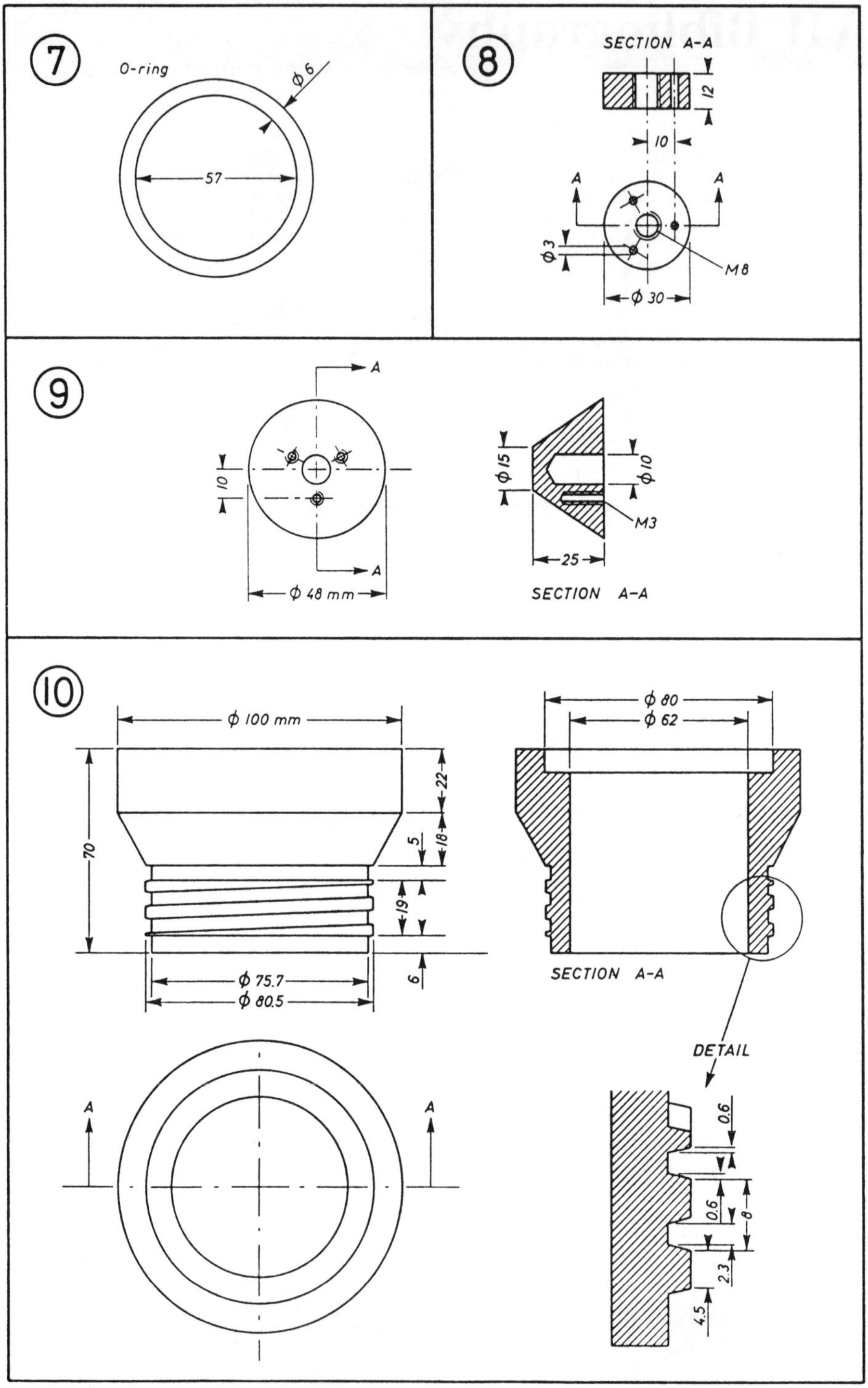

Figure 33. Part IV.

VII Bibliography

Ayer, D.E. & Dynan, W.S. (1988): Simian virus 40 major late promoter: A novel tripartite structure that includes intragenic sequences. Mol. Cell. Biol. *8*: 2021–2033.

Baeuerle, P.A. & Baltimore, D. (1988): Activation of DNA-binding activity in an apparently cytoplasmic precursor of the NF-kB transcription factor. Cell *53*: 211–217.

Bagchi, S., Raychaudhuri, P. & Nevins, J.R. (1989): Phosphorylation-dependent activation of the adenovirus-inducible E2F transcription factor in a cell-free system. Proc. Nat. Acad. Sci. *86*: 4352–4356.

Bazett-Jones, D.P., Yeckel, M. & Gottesfeld, J.M. (1985): Nuclear extracts from ß-globin-synthesizing cells enhance globin transcription *in vitro*. Nature *317*: 824–828.

Bienz, M. & Pelham, H.R. (1986): Heat shock regulatory elements function as an inducible enhancer in the Xenopus hsp 70 gene and when linked to a heterologous promoter. Cell *45*: 753–760.

Bird, A. (1986): CpG-rich islands and the function of DNA methylation. Nature *321*: 209–213.

Blau, H.M. (1988): Hierarchies of regulatory genes may specify mammalian development. Cell *53*: 673–674.

Bohman, D., Keller, W., Dale, T., Schöler, H.R., Tebb, G. & Mattaj, I.W. (1987): A transcription factor which binds to the enhancers of SV 40, immunoglobulin heavy chain, and U2 snRNA genes. Nature *325*: 268–272.

Borelli, E., Hen, R. & Chambon, P. (1984): Adenovirus 2 E1A products repress enhancer-induced stimulation of transcription. Nature *312*: 608–612.

Cedar, H. (1988): DNA methylation and gene activity. Cell *53*: 3–4.

Charnay, P., Treisman, R., Mellon, P., Chao, M., Axel, R. & Maniatis, T. (1984): Differences in human α- and ß-globin gene expression in mouse erythroleukemia cells: the role of intragenic sequences. Cell *38*: 251–263.

Cockell, M., Stevenson, B., Strubin, M., Hagebücle, O. & Wellauer, P.K. (1989): Identification of a cell-specific DNA-binding activity that interacts with a transcriptional activator of genes expressed in the acinar pancreas. Mol. Cell. Biol. *9*: 2464–2476.

Dignam, J.D., Lebovitz, R.M. & Roeder, R.G. (1983): Accurate transcription initiation by RNA polymerase II in a soluble extract from isolated mammalian nuclei. Nuc. Acids Res. *11*: 1475–1489.

Dynan, W.S. (1989): Modularity in promoters and enhancers. Cell *58*: 1–4

Dynan, W.S. & Tjian, R. (1983): Isolation of transcription factors that discriminate between different promoters recognized by RNA polymerase II. Cell *32*: 669–680.

Dynan, W.S. & Tjian, R. (1985): Control of eukaryotic messenger RNA synthesis by sequence-specific DNA-binding proteins. Nature *316*: 774–778.

Fire, A., Samuels, M. & Sharp, P.A. (1984): Interactions between RNA polymerase II, factors, and template leading to accurate transcription. J. Biol. Chem. *259*: 2509–2516.

Fried, M.G. & Crothers, D.M. (1981): Equilibria and kinetics of lac repressor-operator interactions by polyacrylamide gel electrophoresis. Nuc. Acids Res. *9*: 3047–3060.

Freedman, L.P., Yoshinaga, S.K., Vanderbilt, J.N. & Yamamoto, K.R. (1989): *in vitro* transcription enhancement by purified derivatives of the glucocorticoid receptor. Science *245*: 298–301.

Fuller, R.S. & Kornberg, A. (1983): Purified dnaA protein in initiation of replication at the *Escherichia coli* chromosomal origin of replication. Proc. Nat. Acad. Sci. *80*: 5817–5821.

Fuller, R.S., Funnell, B.E. & Kornberg, A. (1984): The dnaA protein complex with the *E. coli* chromosomal replication origin (oriC) and other sites. Cell *38*: 889–900.

Galas, D. & Schmitz, A. (1978): DNAase footprinting: a simple method for the detection of protein-DNA-binding specificity. Nuc. Acids Res. *5*: 3157–3170.

Garner, M.M. & Revzin, A. (1981): A gel electrophoresis method for quantifying the binding of proteins to specific DNA regions. Nuc. Acids Res. *9*: 3047–3060.

Gasser, S.M. & Laemmli, U.K. (1986): Cohabitation of scaffold binding regions with upstream/enhancer elements of three developmentally regulated genes of *D. melanogaster*. Cell *46*: 521– 530.

Gasser, S.M. & Laemmli, U.K. (1987): A glimpse at chromosomal order. Trends Genet. *3*: 16–21.

Gester, T. & Roeder, R.G. (1988): A herpes virus *trans*-activating protein interacts with transcription factor OTF-1 and other cellular proteins. Proc. Nat. Acad. Sci. *85*: 6347–6351.

Goodbourn, S.H., Burnstein, H. & Maniatis, T. (1986): The human beta-interferon gene enhancer is under negative control. Cell *45*: 601–610.

Gorski, K., Carneiro, M. & Schibler, U. (1986): Tissue-specific *in vitro* transcription from the mouse albumin promoter. Cell *47*: 767–776.

Graham, F.L. & Van der Eb, A.J. (1973): A new technique for the assay of infectivity of human adenovirus 5 DNA. Virology *52*: 456–467.

Hamer, D.H. (1986): Metallothionein. Ann. Rev. Biochem. *19*: 209–252.

Heintz, N. & Roeder, R.G. (1982): Transcription of eukaryotic genes in soluble cell-free systems. In "Genetic Engineering" 4, J.K. Setlow & A. Hollaeunder, eds, (New York, Plenum Press), pp. 57–89.

Hogan, B.L.M., Constantini, F. & Lacy, E. (1986): Manipulation of the mouse embryo: a laboratory manual (Cold Spring Harbor Laboratory; Cold Spring Harbor, New York).

Jackson, S.P. & Tjian, R. (1988): O-glycosylation of eukaryotic transcription factors: implications for mechanisms of transcriptional regulation. Cell *55*: 125–133.

Johnson, P.F. & McKnight, S.L. (1989): Eukaryotic transcriptional regulatory proteins. Ann. Rev. Biochem: *58*: 799–839.

Jones, N., Rigby, P. & Ziff, E. (1988): *Trans*-acting protein factors and the regulation of eukaryotic transcription: lessons from studies on DNA tumor viruses. Genes Dev. *2*: 267–281.

Kageyama, R., Merlino, G.T. & Pastan, I. (1989): Nuclear factor ETF specifically stimulates transcription from promoters without a TATA box. J. Biol. Chem. *264*: 15508–15514.

Kalb, V.F. & Bernlohr, R.W. (1977): A new spectrophotometric assay for protein in cell extracts. Anal. Biochem. *82*: 362–371.

Lawn, R.M., Fritsch, E.F., Parker, R.C., Blake, G. & Maniatis, T. (1978): The isolation and characterization of linked δ and ß-globin genes from a cloned library of human DNA. Cell *15*: 1157–1174.

Lewis, M.K. & Burgess, R.R. (1982): Eukaryotic RNA polymerases. In "The enzymes" Vol. XV, P. Boyer, ed. (New York, Academic Press) pp. 109–153.

Lichtsteiner, S., Wuarin, J. & Schibler, U. (1987): The interplay of DNA-binding proteins on the promoter of the mouse albumin gene. Cell *51*: 963–973.

Lichtsteiner, S. & Schibler, U. (1989): A glycosylated liver-specific transcription factor stimulates transcription of the albumin gene. Cell *57*: 1179–1187.

Losick, R. & Pero, J. (1981): Cascades of sigma factors. Cell *25*: 582–584.

Maire, P., Wuarin, J. & Schibler, U. (1989): *in vitro* dissection of the albumin gene promoter reveals positive *cis*-acting elements for tissue-specific and ubiquitous transcripion factors. Science *244*: 343–346.

Maniatis, T., Fritsch, E.F. & Sambrook, J. (1982): "Molecular cloning: A laboratory manual" Cold Spring Harbor Laboratory (Cold Spring Harbor eds.):, N.Y.

Maniatis, T., Goodbourn, S. & Fischer, J.A. (1987): Regulation of inducible and tissue-specific gene expression. Science *236*: 1237–1245.

Maniatis, T., Hardison, R.C., Lacy, E., Lauer, J., O'Connell, C., Quon, D., Sim, G.K. & Efstratiadis, A. (1978): The isolation of structural genes from libraries of eukaryotic DNA. Cell *15*: 687–701.

Manley, J.L., Fire, A., Cano, A., Sharp, P.A. & Gefter, M.L. (1980): DNA-dependent transcription of adenovirus genes in a soluble whole-cell extract. Proc. Nat. Acad. Sci. *77*: 3855–3859.

Mc Knight, S. & Tjian, R. (1986): Transcriptional selectivity of viral genes in mammalian cells. Cell *46*: 795–805.

Melton, D.A., Krieg, P.A., Rebagliati, M.R., Maniatis, T., Zinn, K. & Green, M.R. (1984): Efficient *in vitro* synthesis of biologically active RNA and RNA hybridization probes from plasmids containing a bacteriophage SP6 promoter. Nuc. Acids Res. *12*: 7035–7056.

Messing, J. (1983): New M13 vectors for cloning. Meth. Enzymol. *101*: 20–78.

Mitchell, P.J. & Tjian, R. (1989): Transcriptional regulation in mammalian cells by sequence-specific DNA binding proteins. Science *245*: 371–378.

Nelson, M. & Mc Clelland, M. (1987): The effect of site-specific methylation on restriction-modification enzymes. Nuc. Acids Res. *15*: r219–r230.

Nonet, M., Sweetser, D. & Young, R.A. (1987): Functional redundancy and structural polymorphism in the large subunit of RNA polymerase II. Cell *50*: 909–915.

Parker, C.S. (1989): Transcription factors. Curr. Opinion Cell. Biol. *1*: 512–518.

Parker, C.S. & Topol, J. (1984a): A *Drosophila* RNA polymerase II transcription factor contains a promoter region-specific DNA-binding activity. Cell *36*: 357–369.

Parker, C.S. & Topol, J. (1984b): A *Drosophila* RNA polymerase II transcription factor binds to the regulatory site of an hsp 70 gene. Cell *37*: 273–283.

Parslow, T.G., Jones, S.D., Bond, B. & Yamamoto, K.R. (1987): The immunoglobulin octanucleotide: Independent activity and selective interaction with enhancers. Science *235*: 1498–1501.

Paule, M.R. (1981): Comparative subunit composition of the eukaryotic nuclear RNA polymerases. Trends Biochem. Sci. *6*: 128–131.

Ptashne, M. (1986): A genetic switch. (Cambridge, MA & Palo Alto, CA: Cell press and Blackwell Scientific).

Razin, A. & Riggs, A.D. (1980): DNA methylation and gene function. Science *210*: 604–610.

Saltzman, A.G. & Weinmann, R. (1989): Promoter specificity and modulation of RNA polymerase II transcription. FASEB J. *3*: 1723–1733.

Saluz, H.P. & Jost, J.P. (1990): A laboratory guide for *in vivo* studies of DNA methylation and protein/DNA interactions. Birkhäuser Verlag, Ed., Basel.

Sassone-Corsi, P., Dougherty, J.P., Wasylyk, B. & Chambon, P. (1984): Stimulation of *in vitro* transcription from heterologous promoters by the simian virus 40 enhancer. Proc. Nat. Acad. Sci. *81*: 308–312.

Sawadogo, M. & Roeder, R.G. (1985): Factors involved in specific transcription by human RNA polymerase II: Analysis by a rapid and quantitative *in vitro* assay. Proc. Nat. Acad. Sci. *82*: 4394–4398.

Schibler, U. & Sierra, F. (1987): Alternative promoters in developmental gene expression. Ann. Rev. Genet. *21*: 237–257.

Schöler, H.R. & Gruss, P. (1984): Specific interaction between enhancer-containing molecules and cellular components. Cell *36*: 403–411.

Schöler, H.R. & Gruss, P. (1985): Cell type-specific transcriptional enhancement *in vitro* requires the presence of *trans*-acting factors. EMBO J. *4*: 3005–3013.

Sentenac, A. (1985): Eukaryotic RNA polymerases. Crit. Rev. Biochem. *18*: 31–91.

Serfling, E., Jasin, M. & Schaffner, W. (1985): Enhancers and eukaryotic gene transcription. Trends Genet. *1*: 224–230.

Sergeant, A., Bohmann, D., Zentgraf, H., Weiher, H. & Keller, W. (1984): A transcription enhancer acts *in vitro* over distances of hundreds of base-pairs on both circular and linear templates but not on chromatin reconstituted DNA. J. Mol. Biol. *180*: 577–600.

Shapiro, D.J., Sharp, P.A., Wahli, W.W. & Keller, M.J. (1988): A high-efficiency HeLa cell nuclear transcription extract. DNA *7*: 47–55.

Shortle, D. & Botstein, D. (1983): Directed mutagenesis with sodium bisulfite. Meth. Enzymol. *100*: 457–468.

Sierra, F., Pittet, A.C. & Schibler, U. (1986): Different tissue-specific expression of the amylase gene Amy-1 in mice and rats. Mol. Cell. Biol. *6*: 4067–4076.

Sierra, F., Tamone, F., Mueller, C.R. & Schibler, U. (1990): Liver-specific *in vitro* transcription from the promoter of a rat α2U-globulin gene. Mol. Biol. & Med. *7*: 131–146.

Singh, H., LeBowitz, J.H., Baldwin, A.S. & Sharp, P.A. (1988): Molecular cloning of an enhancer binding protein: isolation by screening of an expression library with a recognition site DNA. Cell *52*: 415–423.

Smale, S.T. & Baltimore, D. (1989): The "initiator" as a transcription control element. Cell *57*: 103–113.

Sorger, P.K. & Pelham H.R.B. (1988): Yeast heat shock factor is an essential DNA-binding protein that exhibits temperature-dependent phosphorylation. Cell *54*: 855–864.

Staudt, L.M., Clerc, R.G., Singh, H., LeBowitz, J.H., Sharp, P.A. & Baltimore, D. (1988): Cloning of a lymphoid-specific cDNA encoding a protein binding the regulatory octamer DNA motif. Science *241*: 577–580.

Sudgen, B. & Keller, W. (1973): Mammalian deoxyribonucleic acid-dependent ribonucleic acid polymerases I. Purification and properties of an α-amanitin-sensitive ribonucleic acid polymerase and stimulatory factors from HeLa and KB cells. J. Biol. Chem. *248*: 3777–3788.

Tamura, T.-A., Aoyama, A., Inoue, T., Miura, M., Okano, H. & Mikoshiba, K. (1989): Tissue-specific *in vitro* transcription from the mouse myelin basic protein promoter. Mol. Cell. Biol. *9*: 3122–3126.

Tsuda, M. & Suzuki, Y. (1981): Faithful transcription initiation of fibroin gene in a homologous cell-free system reveals an enhancing effect of 5' flanking sequences far upstream. Cell *27*: 175–182.

Weil, P.A., Luse, D.S., Segall, J. & Roeder, R.G. (1979): Selective and accurate initiation of transcription at the Ad2 major late promotor in a soluble system dependent on purified RNA polymerase II and DNA. Cell *18*: 469–484.

Weintraub, H. & Groudine, M. (1976): Transcriptionally active and inactive conformations of chromosomal subunits. Science *93*: 848–858.

Wiederrecht, G., Seto, D. & Parker, C.S. (1988): Isolation of the gene encoding the *S. cerevisiae* heat shock transcription factor. Cell *54*: 841–853.

Wigler, M., Pellicer, A., Silverstein, S., Axel, R., Urlaub, G. & Chassin, L. (1979): DNA-mediated transfer of the adenine phosphoribosyl transferase locus into mammalian cells. Proc. Nat. Acad. Sci. *76*: 1373–1376.

Workman, J.L. & Roeder, R. (1987): Binding of transcription factor TFIID to the major late promoter during *in vitro* nucleosome assembly potentiates subsequent initiation by RNA polymerase II. Cell *51*: 613–622.

Wu, C. (1985): An exonuclease protection assay reveals heat-shock elements and TATA box DNA-binding proteins in crude nuclear extracts. Nature *317*: 84–87.

Yamamoto, K.R. (1985): Steroid receptor regulated transcription of specific genes and gene networks. Ann. Rev. Genet. *19*: 209–252.

Young, R. & Davis, R. (1983): Efficient isolation of genes by using antibody probes. Proc. Nat. Acad. Sci. *80*: 1194–1198.

VIII Index